T0338154

Broadband Services

Broadband Services
BUSINESS MODELS AND TECHNOLOGIES FOR COMMUNITY NETWORKS

Edited by

Imrich Chlamtac

CreateNet Research Consortium

Ashwin Gumaste

Fujitsu Laboratories, Dallas, USA

and

Csaba A. Szabó

Budapest University of Technology and Economics, Budapest, Hungary

John Wiley & Sons, Ltd

Other Wiley Editorial Offices

John Wiley & Sons Inc., 111 River Street, Hoboken, NJ 07030, USA

Jossey-Bass, 989 Market Street, San Francisco, CA 94103-1741, USA

Wiley-VCH Verlag GmbH, Boschstr. 12, D-69469 Weinheim, Germany

John Wiley & Sons Australia Ltd, 33 Park Road, Milton, Queensland 4064, Australia

John Wiley & Sons (Asia) Pte Ltd, 2 Clementi Loop #02-01, Jin Xing Distripark, Singapore 129809

John Wiley & Sons Canada Ltd, 22 Worcester Road, Etobicoke, Ontario, Canada M9W 1L1

Library of Congress Cataloguing-in-Publication Data

Broadband services : business models and technologies for community
    networks / edited by Ashwin Gumaste, Imrich Chlamtac, Csaba Szabó.
      p.   cm.
    Includes bibliographical references and index.
    ISBN 0-470-02248-5 (alk. paper)
    1. Broadband communication systems.   2. Electronic villages (Computer networks)
I. Gumaste, Ashwin.   II. Chlamtac, Imrich.   III. Szabó, Csaba, (Csaba A.)
TK5103.4.B767 2005
384–dc22   2004022888

British Library Cataloguing in Publication Data

A catalogue record for this book is available from the British Library

ISBN 0-470-02248-5 (HB)

Typeset in 10/12pt Times by TechBooks, New Delhi, India
Printed and bound in Great Britain by Antony Rowe Ltd, Chippenham, Wiltshire
This book is printed on acid-free paper responsibly manufactured from sustainable forestry
in which at least two trees are planted for each one used for paper production.

Contents

Foreword

It has been shown that there is a correlation between the broadband adoption rate and the gross national products of various countries. Therefore, all industrial countries are racing to ensure that their citizens have the best possible broadband access. In addition to national governments, local governments and municipal organizations are taking steps to increase the broadband access to their populations. Providing broadband access to citizens, communities, and public institutions and developing businesses has become a strategic objective for governments and international organizations worldwide. However, the solution to these problems is not straightforward. Although the appropriate technologies are there, ranging from the almost ubiquitous ADSL, cable modems, to fiber-to-the premises and broadband wireless access, telecommunication and cable companies cannot deploy them as the large costs and very long return of investment makes it prohibitive according to their usual business models.

Providing broadband infrastructure in sparsely populated, geographically challenged areas, usually underserved by service providers, is a very important goal for local governments. A large number of initiatives, under the collective name community networks or municipal fiber networks or metro fiber networks have been launched in North America as well as in Europe. By creating broadband infrastructure in underserved regions, governments can prevent remote communities from "digital divide," and are able to create a climate for economic development, help startups to grow, bring new businesses into the region thus resulting in a healthy climate for economic development.

There is a variety of business models for public entity's participation in building backbone and access networks for communities. According to the one of the most common models, the public administration enters the market, directly or indirectly, as an infrastructure provider, offering basic infrastructure such as ducts and "dark fiber", to other entities which act as service providers for the end users. This model also allows for fostering the competition in the broadband access market by creating an alternative infrastructure to the incumbent telecommunication operator. The right business model depends on national and local telecom regulations, competition regulation and of course economic considerations.

This book, *Broadband Services: Business Models and Technologies for Community Networks*, edited by Imrich Chlamtac, Ashwin Gumaste and Csaba A. Szabó brings together the most important aspects of the problem – technical, legal, regulatory and economic – into one book. The emerging business models and state-of-the-art technologies for broadband networks are covered by an exclusive group of contributing authors from academia, industry and public bodies, including the editors themselves. The editors successfully integrated different contributions into a unified text that provides an integrated view of applications, business considerations and technology aspects of delivering broadband services to business, public and residential users.

There is no similar book currently available in the market. While there exist separate texts on optical or wireless networks, the whole gamut of technologies is not covered in a single book. The technology parts are understandable for the non-technical readers, too. The integrated approach and the coverage of interrelated legal, regulatory and economic issues makes the book really unique. It will be useful for a wide circle of specialists involved in engineering, network planning, business modeling, implementing and operating broadband infrastructures and services, as well as for researchers, academicians, students interested in the various fields.

Raj Jain
Co-Founder and CTO of Nayna Networks, San Jose, CA
Ex-Professor of Computer Science and Engineering,
Ohio State University, Columbus, OH
Fellow of IEEE and Fellow of ACM
http://www.rajjain.com/

About the Editors

Imrich Chlamtac is the President of CreateNet and holds the Distinguished Chair in Telecommunications Professorship at the University of Texas at Dallas, and Bruno Kessler Honorary Professorship from the University of Trento, Italy.

Dr Chlamtac is known as the inventor of the lightpath concept, the basic mechanism for wavelength routing, a key optical technology for WDM networking, and was the first to introduce the fundamental concepts of multihop ad hoc networking. For his various contributions, Dr Chlamtac was elected Fellow of the IEEE, Fellow of the ACM, and, among various awards, received the 2001 ACM Award for Outstanding Contributions to Research on Mobility and the 2002 IEEE Award for Outstanding Technical Contributions to Wireless Personal Communications. He is also the recipient of the MIUR CENECA award in Italy, the Sackler Professorship from Tel Aviv University, the University Professorship at the Budapest University of Technology and Economics, and Fulbright Scholarship.

Dr Chlamtac has published over three hundred and fifty refereed journal, book and conference articles and is the co-author of four books, including the first textbook on LANs entitled *Local Networks: Motivation, Technology and Performance* (1980) and *Wireless and Mobile Network Architectures*, John Wiley & Sons (2002), an IEEE Network Editor's choice, and Amazon.com engineering bestseller.

Dr Chlamtac has contributed to the scientific community as founder and Chair of ACM Sigmobile, and founder and steering committee chair of several leading conferences in networking, including Mobicom, OptiComm, Mobiquitous, Broadnets, WiOpt and others.

Dr Chlamtac also serves as the founding Editor in Chief of the ACM/URSI/Kluwer Wireless Networks (WINET), the ACM/Kluwer Journal on Special Topics in Mobile Networks and Applications (MONET).

Ashwin Gumaste is with Fujitsu Laboratories in the Photonics Networking Laboratory (PNL) group in Richardson, Texas, where his research includes designing next generation photonic and data systems.

Prior to Fujitsu Laboratories, Ashwin received a PhD in Electrical Engineering from the University of Texas at Dallas. Ashwin has previously worked in Fujitsu Network Communications R&D and prior to that with Cisco Systems under the Optical Networking Group.

He has written numerous papers and has over thirty pending US and EU patents. During 1991, Ashwin was awarded the National Talent Search Scholarship in India. His research interests include optical and wireless networking and uncertainty equilibria in social and networking environments. He proposed the first architecture to implement optical burst transport and dynamic lightpath provisioning, called Light-trails, and also proposed the Light-frame framework – a conceptual model for future packet mode optical communication.

Dr Gumaste has authored two books on broadband networks *DWDM Network Designs and Engineering Solutions*, and *First Mile Access Networks and Enabling Technologies.*

Csaba A. Szabó received his PhD from the Budapest University of Technology and Economics (BUTE) and a Doctor of Technical Sciences title from the Hungarian Academy of Sciences. He is a Professor at the Department of Telecommunications of BUTE, a member of the Board of Directors of CreateNet, an international research center based in Trento, Italy, and a member of the Advisory Board of the ICT Graduate School of the University of Trento.

Dr Szabó has achieved recognized research results in multi-access communications and metropolitan area networks. He has published over 50 journal papers, numerous conference presentations and authored or co-authored five monographs. His recent research interests include multimedia communications and planning of community networks.

Dr Szabó has been a member of editorial boards of several leading journals, including *Computer Networks* and *ISDN Systems*. His recent conference organizing activity includes WONS (2004, 2005), OptiComm (2003), the 1st International Workshop on Community Networks and FTTx, where he was a keynote speaker (2003), WiCon and TridentCom (2005). He is a General Co-Chair of the Multimedia Services Access Networks (MSAN) conference and the organizer of the 2nd Workshop on Community Networks.

Dr Szabó is a Senior Member of the IEEE and is listed in the *Marquis Who is Who in the World,* Millenium Edition.

List of Contributors

Enzo Baccarelli
INFO-COM Department
University of Rome 'La Sapienza'
Rome
Italy
enzobac@infocom.uniromal.it

Jim Baller
The Baller Herbst Law Group
2014 P Street
N.W. Suite 200
Washington DC
20036
USA

Anupam Banerjee
Department of Engineering and Public Policy
Carnegie Mellon University
5000 Forbes Avenue
Pittsburgh
PA 15213
USA
anupam_banerjee@cmu.edu

Radim Bartoš
Department of Computer Science
121 Technology Drive Suite 2
Durham
NH 03824
USA
rbartos@cs.unh.edu

Mauro Biagi
INFO-COM Department
University of Rome 'La Sapienza'
Rome
Italy
biagi@infocom.uniromal.it

Raffaele Bruno
IIT-CNR Pisa
Via G. Moruzzi 1
56124
Pisa
Italy

Imrich Chlamtac
CreateNet Research Consortium
Via Solteri 38
38100
Trento
Italy
chlamtac@create-net.it

Marco Conti
IIT-CNR Pisa
Via G. Moruzzi 1
56124
Pisa
Italy

Wael William Diab
Chief Editor, IEEE 802.3ah Task Force (EFM),
Secretary, IEEE 802.3 Working Group (Ethernet),
Technical Leader, Cisco Systems
wdiab@cisco.com

Steven Fulton
InterOperability Laboratory
121 Technology Drive Suite 2
Durham
NH 03824
USA
sfulton@iol.unh.edu

Nasir Ghani
Box 5004
Dept. ECE
Tennessee Technical University
Cookeville
TN 38505
USA

Chaitanya Godsay
Department of Computer Science
121 Technology Drive Suite 2
Durham
NH 03824
USA
cgodsay@iol.unh.edu

Enrico Gregori
IIT-CNR Pisa
Via G. Moruzzi 1
56124
Pisa
Italy
enrico.gregori@iit.cnr.it

Ashwin Gumaste
Fujitsu Laboratories
7421 Frankford Road
631 Dallas
Texas
75252
USA
ashwing@ieee.org

Gareth Hughes
eris@
The European Regional
Information Society Association
19 Rue de Pavie
B-1000 Brussels
Belgium
ghughes@irisi.u-net.com

Roy Isacowitz
NDS Israel Ltd
PO Box 23012
Har Hotzvim
Jerusalem
91235
Israel
risacowitz@ndisrael.com

Arieh Moller
NDS Israel Ltd
PO Box 23012
Har Hotzvim
Jerusalem
91235
Israel
amoller@ndisrael.com

Giovanni Pascuzzi
Universita di Trento
Via Verdi 53
38100
Trento
Italy
pascuzzi@jus.unitn.it

Andrea Rossato
Universita di Trento
Via Verdi 53
38100
Trento
Italy
arossato@istitutocolli.org

Priya Shetty
MS – HR 10
The University of Texas at Dallas
PO Box 830688
Richardson TX
75083 – 0688
USA

Marvin Sirbu
Department of Engineering and Public Policy
Carnegie Mellon University
5000 Forbes Avenue
Pittsburgh
PA 15213
USA
sirbu@cmu.edu

Bill St. Arnaud
CANARIE Inc.
110 O'Connor Street, 4th Floor
Ottawa
Ontario
K1P 5M9
Canada
bill.st.arnaud@canarie.ca

Sean A. Stokes
The Baller Herbst Law Group
2014 P Street
N.W. Suite 200
Washington DC
20036
USA
sstokes@baller.com

Csaba A. Szabó
Budapest University of Technology and Economics
Magyar Tudósok körútja 2
Room: I.L.115
H-1521
Budapest
Hungary
szabó@hit.bme.hu

Scott A. Valcourt
The University of New Hampshire
InterOperability Laboratory
121 Technology Drive Suite 2
Durham
New Hampshire
03824
USA
sav@unh.edu

Acknowledgements

The editors are grateful to the contributing authors for their submissions and willingness to revise their materials to better fit into the framework of the book. Without their cooperation, this project would not have been possible.

The greater part of the book is based on the contributions made by selected presenters at the 1st International Workshop on Community Networks and FTTx, held in Dallas, Texas, in October 2003. We would like to thank them for the hard work of converting their presentations into book chapters, and responding to the editors' numerous requests. The remaining chapters were written by authors who were asked by the editors to contribute with specific topics not covered at the workshop, and which also satisfied our reviewers' suggestions. This way we hope that the book has become a well-integrated and consistent treatment of the most important aspects of providing broadband services. If this is the case, it is due to the efforts of this second group of authors too, who contributed with original and quality material.

Last, but not least, we would like to thank the wonderful editorial staff of John Wiley Europe, who were always helpful in guiding the editors through the long process of preparing this book.

1

Introduction

Imrich Chlamtac[2], Ashwin Gumaste[1] and Csaba A. Szabó[3]

[1]*Fujitsu Laboratories, Dallas, USA*
[2]*CreateNet Research Consortium, USA*
[3]*Budapest University of Technology and Economics, Budapest, Hungary*

1.1 What is broadband?

This widely used term has different meanings. Historically, by broadband we meant cable TV to homes. While CATV is still an option for providing broadband services to homes, the generally accepted meaning of 'broadband' is wider and less specific. It means a high-speed, always-on connection. The meaning of 'high-speed' will definitely change over time, these days it means access rates of about 256 kbps and higher, at least in the downstream direction, for web downloads (definitely higher than the ISDN basic rate, which is 64 kbps or 128 kbps when using one channel or two channels, respectively). The always-on feature is not fully exploited yet but is becoming increasingly important. The Asymmetric Digital Subscriber Line (ADSL) technology and service, widely offered by telecommunication service providers, satisfies the above criteria and is the most widely used broadband service at present. It is a perfect solution for average Internet usage, given its typical 128 kbps upstream and 512 kbps downstream data rates.

A more general, and at the same time dynamic (and therefore future-proof), definition of broadband is: 'the access link performance should not be the limiting factor in a user's capability for running today's applications' [1]. This also refers to the interaction between network technology and applications: network performance has to support today's application needs and will have to be improved to meet the new bandwidth and quality of service requirements when new applications arise.

There were 63 million broadband subscribers worldwide at the beginning of 2003, and while this number is growing, it still represents a small fraction of the total 1.13 billion fixed line subscribers, and a slightly higher number of mobile users [2]. Moreover, there are huge differences among different countries, even within the OECD group. Figure 1.1 shows the top 15 countries, indicating the distribution of the number of broadband subscribers per

Broadband Services: Business Models and Technologies for Community Networks. Edited by I. Chlamtac,
A. Gumaste and C. Szabó © 2005 John Wiley & Sons, Ltd. ISBN 0-470-02248-5.

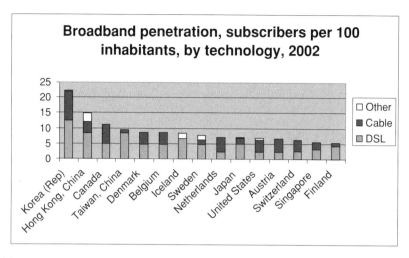

Figure 1.1   Broadband penetration in the top 15 countries. Reproduced with the kind permission of ITU

100 inhabitants over different technologies. DSL is dominating, the share of CATV differs from country to country and 'other' technologies usually represent a small fraction. This latter category includes fiber-to-the-home (FTTH), wireless LANs (WLANs) and others. The number of really high-speed access solutions such as FTTx (where 'x' can be the home or the curbside, just to mention two typical cases) is growing rapidly. According to the latest statistics published at the 2003 FTTH Annual Conference, there were 800 000 FTTH subscribers at the beginning of 2003, again, with a highly uneven distribution: Asia is represented by 64 %, Europe/Middle East/Africa by 24 % and North America by 12 % [3].

Since the technology cannot be the (only) driving force, let's answer the question users frequently ask: what benefits can they expect from broadband? Speed matters, but not for all services the users want to use. First of all, users do not understand (and do not have to understand) what 256 kbps or any other data rate means. What they need to be told is what these figures will bring them, for instance in terms of download times, when browsing the Internet. The next message could be that entertainment services will be offered via broadband, some of which are currently available only through traditional terrestrial wireless or satellite broadcast (radio and TV programs), and some of which are not available currently via traditional video broadcast and distribution networks (video-on-demand, pay-per-view, personal video recorder). Several large telecom operators are in the process of introducing video over DSL services (France Telecom, BellSouth). For them, the objective of entering into the content provision business ('triple play' as they call it – voice/data/video services) is to compensate for the decrease of revenues from fixed line services and to enable further growth. Thus, a convergence in technology and services results also in competition in different, previously separate areas (access, long-distance carriers, cable operators, content providers) and in convergence in the service provider market. Customers will benefit from this competition; for example, they will deal, in most cases, with a single service provider whom they pay for the applications they are using, instead of being billed for the content, access, Internet and other components separately.

Coming back to broadband technologies, let's ask the question: is ADSL enough? Or, more generally, can the DSL type solutions, based on the use of the traditional copper subscriber loops, support broadband services? The answer is yes, xDSL (where 'x' stands for enhanced versions of DSL such as, for example, HDSL) offers enough data rate even for broadcast quality TV delivery. But there is another issue to deal with: the competition. Local loops belong to incumbent telcos and although they are obliged to offer them to competitive telcos, the local loop unbundling has not been a success so far. Therefore, new telecom service providers often build parallel infrastructures consisting of optical core and distribution networks plus a variety of access networks, including FTTH and wireless LANs.

Public participation in building broadband infrastructure and fostering the development of broadband services is often advisable for several reasons. First, there are rural, sparsely populated or geographically challenged areas where it is not possible to build telecommunications solely based on market conditions. To put it simply, a for-profit telecom company is not interested in, or more exactly, cannot afford, building an infrastructure if the return of investment is too long. To solve the needs in rural areas, some form of public intervention is needed. For local governments it is often a straightforward step to become – directly or indirectly – a telecom infrastructure operator, since their high-speed networks interconnecting public institutions are already in place and are being upgraded. In addition to owning networks, there are several other models of how the public can participate in telecommunication development, thus bridging the 'digital divide'. This way, public administrations can ensure that society-related objectives are met and that the citizens can enjoy the benefits from broadband. To mention some examples: citizens will be able to arrange their administration affairs without travel, or receive medical treatment by the best specialists or be examined by the latest technology diagnostic equipment without the need for traveling to remote health care providers.

Recognizing the importance of broadband, national governments and, in Europe, the European Union have launched broadband initiatives, programs and projects. The eEurope2005 Action Plan of the European Community bears the title: 'An Information Society for All'. It focuses on achieving widespread availability and use of broadband. Two groups of actions have been defined: one regarding stimulating service, applications and content, and the other about broadband infrastructure and security. The Action Plan contains specific actions for the main application areas of e-government, e-health, e-learning and e-commerce. For example, the e-government actions include broadband connection to every public institution, introduction of interactive public services, public procurement procedures and creation of Public Internet Access Points (PIAPs) [4]. The guidelines for working out national programs include technology and platform neutrality, geographical focus (on areas that might otherwise be neglected), and open access (creating infrastructure that will be open to all operators and users).

In North America, Canada is very advanced in terms of networking and IT applications in all areas, leading in online governmental services, computerization and interconnection of schools (100 % of schools and libraries are connected), and 30 % share of broadband access. The nationwide connectivity and development of broadband Internet and next generation services are enabled by CA*net4, Canada's next generation optical Internet backbone, developed and managed by CANARIE, Canada's Internet development organization [5].

1.2 The objectives of this book

This book is intended to provide an integrated view of applications, business considerations and technological aspects of delivering broadband services to business, public and residential users.

Given the prevailing assumptions that the demand for broadband services is growing and that technology deployments will further increase the opportunities and facilitate broadband access, this book covers not only broadband technologies, but also the state of the art in terms of user needs for broadband applications and services and the social and legal setting for its deployment and expected evolution.

The book starts by introducing the various applications and services available today and those expected to drive a growing demand for broadband solutions in the future.

Given that the deployment of these services often involves regulatory issues, we next analyze the legal environment that determines the playing field for traditional telcos, cable companies and new entrants, as well as opportunities and limitations for efforts initiated and controlled by the public sector, the latter often referred to as municipality/community networks.

Business models for the leading approaches are introduced, analyzed and compared. We look at the business limitations of the telecoms when planning broadband services and why a local government initiated/supported/controlled approach can be feasible from a business point of view. In the context of a local government or community network solution, we discuss the society-related goals, such as serving sparsely populated/rural areas, promoting e-government services, and, in general, contributing to bridging the 'digital divide'.

The technology part of the book provides an in-depth treatment of existing and emerging solutions for delivering broadband services. The emphasis will be on the analysis of access solutions, the most challenging and important technology area, given the wide range of technologies available and the complexity of the related economical/business aspects. Stability and future-proofness of access technologies is of primary importance. We thus pay special attention to the current standardization situation. We complete the technology design part by addressing state-of-the-art solutions for building the backbone network and finally, by presenting guidelines for technology selection.

The book concludes with a presentation and analysis of a number of case studies, carefully selected from several hundreds of community/metro fiber/municipality networks in both North America and Europe. Through these case studies, the book demonstrates how the business models and technology designs presented in earlier chapters can be used and implemented in different environments.

1.3 Book outline

In *Part One*, *Applications and Services*, Chapters 2 and 3 deal with the applications and services the users need.

In *Part Two*, *Business Models*, Chapters 4 through 8 focus on the legal, regulatory and economic aspects of the development of broadband community network infrastructures and services.

In *Part Three*, *Technology*, Chapters 9 through 14 are devoted to the most important access and metro area technologies that are effective, standard-based and future-proof for delivery of broadband services to the users.

In *Part Four*, *Case Studies*, Chapters 15 and 16 demonstrate some characteristic examples of community area networks in North America and Europe.

Chapter 2: Broadband home/entertainment services

As the title implies, this chapter deals with services wanted by residential users. We show that broadband is in the process of redefining the concept of 'home entertainment.' Then, the most important current and emerging applications, their driving forces and potential markets, as well as the respective industry players, are discussed. We also point out that the genesis of broadband in the Internet culture of free content could severely limit its commercial expansion, and to reach its true commercial potential it will need to respect copyright and generate a culture of payment.

Chapter 3: Applications and services to meet society-related needs

This chapter gives an overview of a group of applications that are important for society and that can be greatly improved by broadband delivery to the users. This group includes applications and services falling into the broad categories of e-learning, e-health and e-government. We show that advanced learning, telemedicine and e-government services, made possible by broadband, bring real benefits to the citizens.

Chapter 4: Key legal and regulatory issues affecting community broadband projects in the United States

Municipal or public sector providers of broadband communications networks in the United States may be subject to various federal, state and local requirements, depending on where they are located, what services they provide and how they structure their activities. In this chapter, the key legal and regulatory issues are reviewed, accompanied by numerous case descriptions.

Chapter 5: European telecommunication law and community networks

The aim of this chapter is to provide a brief overview of the general principles governing the European telecommunications regulatory framework. The legal requirements to provide community network services in the liberalized European market are analyzed. The New Regulatory Framework (NRF) adopted by the EU consists of a set of directives which must be implemented by Member States. The Italian implementation and a case study of community network services are briefly presented.

Chapter 6: Models for public sector involvement in regional and local broadband projects

The choice of model for public participation depends on local circumstances, the legal situation of the country concerned, and also, to a large extent, on the will of the local or regional community. This chapter provides an overview of different models for deploying broadband and attempts to highlight the advantages and disadvantages of each approach. After discussing the models, the different forms of public–private partnerships are outlined. Some legal and economic considerations related to the models are also provided.

Chapter 7: Customer-owned and municipal fiber networks

Dark fiber is considered to be technologically neutral and, in many cases, presents the least problematic construction from a legal/regulatory point of view. We show that customer-owned dark fiber can be the right solution for school boards, municipalities and other public institutions, as well as for businesses. This chapter focuses on planning, implementation and operational issues of dark fiber networks. Different forms of ownership, operation and management of dark fiber networks are discussed and a cost analysis shows the attractiveness of dark fiber from a financial point of view.

Chapter 8: Towards a technologically and competitively neutral fiber-to-the-home (FTTH) infrastructure

The engineering economics of four different FTTH network architectures are considered in this chapter. Different models for competition in the FTTH industry are defined. Results from the engineering cost models of these architectures in three different deployment scenarios are used to comment on the implications that network architecture has for competition. We show that the lowest cost FTTH architecture supports different models of FTTH competition. We conclude with a discussion on issues in FTTH industry structure.

Chapter 9: Backbone optical network design for community networks

Design and optimization of community networks is a critical task if one wants to achieve stable, convergent and scalable networks. A typical community network can be considered as a network of networks spanning the entire community and giving access to end users, customers, enterprises and consumers. It can be hierarchically classified into two or three sectors – the first/last mile area, the collector/metro access area and the backbone or the metro core area. The last two aspects are treated together, as the only difference between the two is the network size. The first mile, on the other hand, consists of a set of varied and diverse technologies.

Chapter 10: A comparison of the current state of DSL technologies

In the mid 1990s, the vision for an asynchronous last mile link to deliver video services over existing voice-grade cable drove the early investment in Asymmetric Digital Subscriber Line (ADSL). With traffic patterns of high-speed Internet access, ADSL and its implementation began with a completely different market than originally envisioned. In the next generation usage of the DSLs, the convergence of voice, video and data is a driving factor. In this chapter, we highlight the major features and benefits of ADSL, SHDSL and VDSL technologies. We address implementation issues, as well as emerging standardization efforts affecting the DSL technologies.

Chapter 11: Fiber in the last mile

Despite the seemingly high cost structure, fiber solutions are fast becoming prominent in the last mile, for two reasons. Primarily, the other three technological solutions (copper, wireless and power line) cannot meet the emerging needs of high-bandwidth services like video-on-demand, and hence, fiber solutions are the best alternative for providing high-bandwidth ubiquitous services. Secondly, the cost of fiber solutions has considerably declined due to maturity in the associated technology.

In this chapter, we describe the topological model and architecture for PON systems, we discuss the different types of passive optical network, namely APON, GPON, EPON and WDM PON, and the components in PON systems.

Chapter 12: Ethernet in the first mile

Ethernet in the first mile, frequently called EFM, combines a number of new access technologies with the ubiquitous Ethernet to provide a cost-effective, standard-based, simple solution for the access part of metropolitan networks. EFM introduces a couple of copper solutions and a set of passive optical solutions, as well as management and environmental capabilities. This chapter summarizes the results of the standardization effort within the IEEE 802.3ah Working Group, which resulted in an approved standard in 2004.

Chapter 13: DOCSIS as a foundation for residential and commercial community networking over hybrid fiber coax

Cable TV distribution systems play an important role in delivering broadband services. CATV systems distribute RF signals from a central location, called the 'headend', to the end sub-scriber nodes. The current cable distribution systems include a combination of coaxial cable with fiber optic cable, and are commonly called hybrid fiber coax (HFC) networks. New cable-based services, such as Internet access and digital telephone, provide interactive connectivity to the end user, so the cable system needs to be bidirectional. This chapter deals with DOCSIS, a technical standard developed originally by CableLabs, which defines a transmission mechanism to provide data network connectivity through an HFC cable system.

Chapter 14: Broadband wireless access networks: a roadmap on emerging trends and standards

Broadband wireless communication is expected to expand social and economic benefits arising from entertainment, e-commerce, teleworking, e-learning and e-government. It may also be the desired bridge toward the ubiquitous and pervasive utilization of the Internet platform. The aim of this chapter is twofold. First, it provides a snapshot of current technologies, standards and regulation policies involved with the Broadband Wireless Access (BWA) paradigm. Secondly, it explores some emerging BWA trends, future BWA potential benefits, projected demands, expected progress and challenges for the multiple (often contrasting) technological options envisaged for providing BWA to the end users. Future trends, such as Wireless Personal Area Networks, Wireless Local Area Networks, Wireless Metropolitan Area Networks and satellite-based systems, are also presented and compared.

Chapter 15: Community case studies in North America

In this chapter, we show how the technology, business and society-related principles of community networks are implemented in practice. We focus on implementation of these concepts by studying some deployment case studies in North America, primarily those that are unique and have different solutions.

Chapter 16: European broadband initiatives with public participation

This chapter is a selection of five characteristic examples of broadband projects created by some form of public participation in different countries in Europe. The case studies differ by the business models, products and services provided, models for public participation and the underlying technology. The five case studies are: the Stockholm municipality's dark fiber network, Fastweb, a fully-fledged multimedia service provider in Milan, the infrastructure created by Endesa in the Spanish city Zaragoza, based on power line communication technology, the satellite-based solution successfully tested in the South West Region of Ireland, and finally, a special form of public participation, the broadband demand aggregation, implemented in the UK.

References

1. Committee on Broadband Last Mile Technology (2003), *Broadband. Bringing home the bits*, Computer Science and Telecommunications Board, National Research Council, National, Academy Press, Washington, DC.
2. International Telecommunication Union (2003), *Birth of broadband*, ITU Internet Reports, September.
3. Whitman, B.(2003), International FTTH Deployment, FTTH Conference 2003, October 7, 2003.
4. EC (2002), *eEurope 2005: An Information Society for All*, Action Plan, Seville, June 21–22, 2002.
5. www.canarie.ca

Part One

Applications and Services

2

Broadband Home/Entertainment Services

Arieh Moller and Roy Isacowitz
NDS, Jerusalem, Israel

2.1 Introduction

For the early adapters, broadband was an end in itself. High-speed surfing was reason enough to put up with the dodgy installations and shoddy service that characterized the early days of the technology. But for the mass of users, broadband comes into its own as a means to an end. And that end is primarily home entertainment – though in its widest sense. Twenty years ago, few would have regarded sitting at a keyboard as fitting the definition of entertainment, yet that is precisely how millions of teenagers spend their leisure time today. Surfing, chat rooms and instant messaging are only some of the terms that have joined the lexicon of entertainment in the past decade.

It is clear that the Internet has had a profound influence on how people spend their leisure time. If entertainment is 'something that amuses, pleases or diverts' [1], the Internet is probably the most significant development on the entertainment scene since moving pictures. But the Internet and movies are also very different. Movies are a *form* of entertainment; the Internet is its *agent*. In modern jargon, the Internet 'empowers' a wide range of entertainment types, many of them unknown before the advent of the public network.

Broadband, in turn, empowers the Internet. It is changing – and will continue to change – the types of content (or media) that are available on the Internet. It will bring rich media, which were previously the preserve of professional environments, into the home. Our living rooms and dens will be transformed into concert halls, movie theaters and games arcades.

So, where the Internet transformed the definition of entertainment, broadband is in the process of reshaping our concept of the *home*. Between them, they are redefining the popular perception of *home entertainment*.

Broadband Services: Business Models and Technologies for Community Networks. Edited by I. Chlamtac, A. Gumaste and C. Szabó © 2005 John Wiley & Sons, Ltd. ISBN 0-470-02248-5.

There is an additional piece in the home entertainment puzzle: the home network. With broadband, the concept of networking home electronic devices for the first time made sense. That led to the emergence and rapid uptake of wireless Ethernet (the 802.11 standard), starting in 2003. With affordable, user-friendly and noninvasive networking mixed in with the Internet and broadband access, home entertainment was ready for a new era.

2003 was a watershed year in the adoption of broadband and wireless technologies. Prior to 2003, the use of leisure time in the home had not changed much since the intrusion of the Internet in the mid to late 1990s, with its chat rooms and infinite surfing possibilities. Even then, the mainstays of home entertainment did not change much. We still watched movies on the TV and listened to music on the stereo. Games were typically limited to one or two players on a PC or a dedicated games console. That began to change in 2003 and will continue to change through the rest of the decade. The distinctions between home electronics and computers, which have persisted stubbornly in the face of regular threats of convergence, have finally begun to disappear. Convergence is becoming a reality, thanks to the twin influences of broadband and wireless networking. By the end of the decade, a DVD without a network connection will be as rare a beast as the dodo or the VCR (remember them?) Perhaps DVDs themselves will be history.

2.2 Current broadband and home entertainment markets

Comparing broadband and dial-up usage is not easy, if only because broadband is measured in lines, while dial-up is measured in users. Nielsen/NetRatings, for example, counts 'all members, two years of age or older, of households that currently have access to the Internet from a personal computer' when determining its total Internet universe. Global Reach, a Web-based marketing communications company, estimated the total Web universe at the end of 2003 at 680 million users. Compare that with the 100 million lines estimated by Point-Topic at the end of 2003. If the average broadband-enabled home consisted of four people above the age of two, those 100 million lines would clearly have accounted for over half the Internet users at the time.

Whichever way you look at it, the growth of broadband has been spectacular. In less time than it takes to recycle the Olympics, broadband has joined the mainstream in the developed world, with the top ten countries in terms of broadband lines reading like the participants in a Group of Eight conference to which China has been invited. What lies behind this surge?

The first reasons are accessibility and price. If a new technology is easy to implement and not much more expensive than the incumbent technology, consumers in the wealthy countries do not seem to need much persuasion to swap over. Both DSL and cable broadband use wiring that already exists in the home and price wars have driven down prices until abandoning dial-up for broadband almost seems like a no-brainer. In the US, ADSL prices dropped by some 25 % in 2003 alone. Note that DSL – Digital Subscriber Line – refers to a frequency division technology implemented on a conventional – copper – subscriber line, whereby lower frequency phone signals are delivered to the home unaltered, while digital signals traverse the phone line at higher frequencies for delivery to end stations such as PCs. 'Asymmetric' in ADSL refers to the fact that the downstream (to the user) channel's data rate is different – higher – than that of the upstream (to the network) channel. ADSL is therefore well suited to general Internet use.

The lure of a new technology may be enough to hook the geeks and early adopters, but it is seldom sufficient to turn it into a mass-market phenomenon. There has to be some utility – or at least enormous fun – in what is offered. With broadband, that utility is access speed and, to an even greater degree, a form of ubiquity. Broadband is always-on; available at a key stroke. That is manna from heaven for most Internet users.

Price, ease of use and utility are not to be sneered at. Many products build a market around just one of them. But nor are they the stuff of which greatness is made. The Internet – truly an epochal technology – created demand where none existed before. Broadband is different. Like most other products, it provides new and better ways of satisfying a demand that already exists.

Video, music and games have been around and in demand for a long time, but it was the peer-to-peer, file sharing sites on the Web – the Napsters and the Kazaas – that stoked the appetite for rich media on the Internet. That, in turn, created the demand for a faster and more efficient means of Internet delivery. Broadband provided that means.

The irony of broadband's genesis in the peer-to-peer netherworld (not as a technology but as a popular service) is that the very thing that initially fanned the demand for broadband could be the thing that chokes it as well. File swapping was born into the Internet culture of free content. Breaking copyright was the *raison d'etre* of Napster (in its first incarnation), Kazaa, Gnutella and others, not to mention the thrill it gave their many devoted users. Paying for music or movies is anathema to them.

But if broadband is to bring about a true metamorphosis in home entertainment, it will have to be as the carrier of prime and timely multimedia (such as movies, music and games). Prime content costs money to create, package and distribute. The owners of the content – music publishers, film studios, games studios and the like – have no intention of making it available for free on the Internet. Even if users were prepared to pay for the content, the owners would be loath to make it available on a public network in digital form for as long as leakage could adversely affect sales in traditional channels (movie theaters, DVDs, video rental and the like.)

The vexing issues of copyright, payment on the Internet and content security are dealt with briefly at the end of this chapter. Suffice to say at this point that the clash between the Internet culture of free content and the determination of the content owners to receive fair remuneration could well determine the long-term commercial success of broadband. Without payment, broadband will be no more than a technology; a faster and more convenient way of accessing the Internet. It will be of little interest as a social and cultural phenomenon.

2.2.1 Television

If 2003 was the year of broadband and wireless in the networking arena, it was the year of the personal (or digital) video recorder when it came to TV. It may have taken the better part of five years for PVRs to make their mark, but when they did, it was nothing short of spectacular. Cable networks and satellite service providers rushed to market with their own versions of digital recorders, almost all of them integrated with the signal receiver.

A PVR or DVR is a device that allows a user to record broadcast digital video or audio to a hard disk, and play the content back when they want to consume this content. This allows for 'time shifting' of the broadcast content with respect to the original broadcast.

The PVR boom overshadowed other TV developments, such as the continuing transition to digital TV and the halting attempts to jumpstart Interactive TV. Only BSkyB in the UK made a success of its interactive endeavors. A Yankee Group research paper published in November

2003 [2] was entitled *Europe's Connected Consumers Love the Internet, but their Passion for TV Still Burns.* Not the catchiest of titles, perhaps, but accurate nonetheless. As the paper went on to say, 'most consumers still see TV and related devices as their main source for paid content; they show a marked reluctance to pay for content from the Internet.'

The implication is clear. If broadband is to achieve its full potential, it will need to somehow co-opt the TV – and not by delivering broadcast TV and video on demand to the PC. Numerous studies have shown that the great majority of TV viewers want to continue watching it on a TV screen.

There is more than one way of providing broadband TV. Microsoft and the PC manufacturers are betting that it will be via the PC, which is the device at the residential end of virtually every broadband line. Thus, Microsoft rolled out a variety of products in late 2003 and 2004, all aimed at extending the reach of the PC (equipped, of course, with Microsoft Windows XP Media Center Edition) until it rubs shoulders with the TV. But broadband is not synonymous with the Internet and it need not terminate at the PC. Both twisted pairs (in the case of DSL) and coax lines (in the case of cable) were originally intended for other purposes and today both can simultaneously carry voice, data and high-quality video. Telcos around the world are beginning to connect ADSL lines to TV set-top boxes in an attempt to replace at least some of the revenue lost to competitive voice services with live TV and video-on-demand.

As the Yankee research paper correctly pointed out, 'satellite and cable TV companies are well established as suppliers of multichannel TV services.' It is going to be a long, hard slog for IP network operators to dislodge them, even though the logic of providing full voice, data and video services over IP is inescapable.

As for the prospects of convergence, the Yankee document maintained that 'the TV experience is now being enhanced by a range of equipment, such as DVDs, home cinemas and personal video recorders (PVRs), which are not connected to the Internet.' Conventional wisdom has it that it is only a matter of time until all these devices are connected by a (probably wireless) home network, with at least one of its constituent parts having a broadband connection to the Internet.

The feature movie for the rest of the decade will be 'Consumer Electronics Meets the Broadband Internet' – and we could be in for some surprises.

2.2.2 *Mobile*

Currently, the types of entertainment that exist for mobile devices are limited by the general capabilities of the particular wireless network used.

Within the home, wireless networks are generally based on the 802.11 standard. These networks are limited by the lack of good content protection standards, QoS problems and interference from other devices, such as cordless telephones and Bluetooth devices. Due to these limitations, there are few, if any, dedicated entertainment services targeted specifically to this medium.

In addition to home Wi-Fi networks, there are also entertainment services provided to mobile devices over the existing cellular networks. These services are today limited by the low bandwidth of the networks, and are generally restricted to services such as ring-tone downloads or short video clips. A 2004 study from Juniper Research [3] found that the market for mobile ring-tones had already peaked and was showing signs of saturation. Worth just over $1 bn (worldwide) in 2003, the sector was forecast to dwindle to around $490 m by 2008.

2.2.3 Consumer products

While video promises to be the highlight of the broadband home, it was not the focus during the technology's early days. That honor went to music and games, both of which changed dramatically as a result of the partnership.

By January 2004, sales of music over the Internet (i.e. legal music downloads) had become the second most popular singles format in the UK, according to the Official Charts Company [4]. More than 150 000 legal downloads were reported the previous month. That was dwarfed by the over 340 000 CD singles sold in the same period – but even that showed a sharp decline.

Sales of CD singles fell by 30 % in the UK in 2003 and Emmanuel Legrand, Bureau Chief for Billboard Europe told the BBC's *The Music Biz* program that he expected CD singles to entirely disappear from the shelves in the US within the next three years [5]. Both developments are attributable primarily to the spread of broadband.

Taking their place are legal download websites, such as Apple's iTunes, which was estimated to account for as much as 50 % of all downloads in early 2004. By allowing people to choose individual tracks, rather than full albums, iTunes undercut the arguments of illegal downloaders that albums were too expensive for the one or two tracks that they really wanted.

Another big winner was the new generation of compact, high-density digital music players such as the iPod, also from Apple. Lighter than two CDs and capable of holding up to 10 000 songs (it comes in 15 GB, 20 GB and 40 GB models), the iPod combined with iTunes to create the new broadband music paradigm, courtesy of Apple.

In fact, broadband and its associated technologies were propagating so fast by early 2004 that even the mighty DVD showed signs of early obsolescence. The DVD's rapid rise during the 1990s was crowned by the acknowledgement by several Hollywood studios in late 2003 that most new movies were actually bringing in more revenue from DVD sales than from their release in movie cinemas. Yet, in March 2004, research by the In-Stat/MDR research group indicated that DVD growth was threatened by the rise of two broadband-based technologies and was unlikely to continue beyond another year or two [6]. Those two technologies were home digital recording technologies and home broadband connections. By the end of 2008, In-Stat/MDR projected, over 270 million consumers would have equipment to make their own media, and over 247 million would be connected to broadband connections. The impact on packaged entertainment goods would be great, the report predicted.

Unlike the music and DVD industries, which saw broadband as a lot more of a threat than an opportunity, the games industry – and in particular the manufacturers of games consoles – saw only dollar signs. Juniper Research reported in March 2004 [3] that the new generation of consoles, scheduled for release early the following year, would be no less than 'intelligent home media centers.' Once again, it was broadband – described as 'high-speed Internet gaming and wireless connectivity between machines and handhelds' – that had precipitated the leap. Juniper predicted that online gaming from video-game consoles would reach almost 28 m regular users by 2008, while next generation handheld hardware revenues were forecast to reach $25 bn in the same period.

2.3 What drives broadband?

In order to figure out what drives broadband, it is necessary first to understand its role in the home. That is more difficult than it sounds. At its most basic, broadband is simply a bandwidth-rich connection to the public Internet. But that, on its own, is of no use to anybody.

The utility of the broadband connection is defined by the service that it provides, or, to put it another way, by the device that is connected to the home end of the broadband pipe. Most people would tend to assume that the device is a PC, because that is the way it has always been. But it takes only a slight leap of the imagination to picture the broadband pipe running directly into a TV (probably via a set-top box) or a games console, or even a stereo set. And it is clear that the service provided by the broadband connection would be very different in each case.

It is likely that broadband in the home will follow a dual track, at least for the conceivable future. On the one hand, it will terminate at the PC, where it will facilitate a wide variety of rich media and games. The drivers for that service will be the ones we know well from the Internet – PC ownership, Internet and broadband usage, the availability of popular games and other content and the comfort of well-known brands.

The second, and later, track will be the spread of broadband beyond the PC into the home, first via the intermediation of the PC and then directly into a variety of consumer devices, probably by means of a home network. The services available will, for the most part, mirror those available on the PC, but they will be rendered on dedicated, high-quality players. Broadcast TV and video-on-demand will be played out on the TV, multiplayer games on the games console, and so on.

Given the gap between TV and PC penetration (118 million households versus some 70 million households in the US), the extension of broadband to the TV would give it a massive boost. Other drivers are the availability of game consoles and other non-PC broadband devices, including digital cameras, PDAs (personal digital assistants) and mobile phones.

There is an interesting alternative to the above scenario. Rather than the PC slowly giving way to a variety of consumer devices as the broadband conduit into the home, the possibility exists that the PC will itself become the multifunctional consumer device. After all, it already plays music, displays movies and acts as a telephone for VoIP. It is a prospect that horrifies the computer electronics manufacturers, but cannot be discounted.

Another alternative is what the Yankee Group calls the Home Network Media Node [7]. These devices, also called connected media adapters, enable consumers to play back PC-based content through their television and stereo systems. They first began appearing on shelves in the US in late 2003 and Yankee projects that they will penetrate 18.5 million US homes by 2007.

It is clear from the above that broadband home entertainment is likely to go through several evolutionary phases (see Table 2.1).

The migration from the PC to consumer entertainment devices is the factor that will most influence the spread of broadband in the long term. So long as broadband is tethered to the PC, it will be regarded, first and foremost, as a technology – and its reach will inevitably bump

Table 2.1   Broadband home entertainment: evolutionary phases

Phase 1	Tethered to PC	Fast Internet; music downloads; multiplayer games
Phase 2	PC as intermediary to consumer devices	Music downloads to handheld devices;
Phase 3	Consumer devices with broadband connection	Multichannel TV and VoD; broadband games consoles
Phase 4	Broadband in every room (wired or wireless)	Remote device management

up against the technophobia of the mass market. Only when its evolution enters Phase 3 (see Table 2.1) will broadband be regarded as a true mass consumer phenomenon. To be truly ubiquitous, broadband must become as unconscious a presence in the home as electricity or the telephone – or cable in North America. No-one hesitates to turn on an electric light because he or she is intimidated by the technology that drives it. In the public perception, electricity is a given – not a technology – and therefore there is no knowledge barrier to its use. Likewise cable, in the countries in which it is prevalent. In other words, broadband could become a utility, in the sense that water, electricity and cooking gas are utilities: cheap, ubiquitous, idiot-proof and essential for the conduct of a modern life.

2.4 Future broadband entertainment services

As noted at the start of this chapter, the Internet has given rise to a variety of activities that, in the past, would not have been identified as entertainment. These include email, chatting, instant messaging and plain old surfing. Not only is the Internet a new entertainment category, it has fast become the most popular. In a Knowledge Networks/Statistical Research study [8] conducted in April 2002, more children aged 8–17 said that they would pick the Internet rather than television if they could have access to only one medium.

Alcatel, a leading supplier of broadband network technologies, has identified a core group of three applications, which, it says, are the prime drivers of broadband growth:

- television – personalization of TV, particularly via video-on-demand (VOD), personal video recording (PVR) and network-based PVR (nPVR);
- consumer electronics products, particularly games consoles;
- broadband Internet.

Of the above applications, we have already mentioned PVR. nPVR is the provisioning of PVR type functionality, i.e. video recording, on a network-based device. Video-on-demand is a method for providing video services to viewers under the viewers' control. This permits them to choose what they want to view and when. VOD often includes the ability to stop, rewind, fast forward and pause during viewing.

In addition, Alcatel has come up with a second tier of applications, which, it says, are not necessarily drivers of broadband, but will provide a significant source of revenue and margin for the service provider. These are:

- online music;
- gambling;
- information services;
- voting;
- T-commerce.

In mid 2002, Alcatel commissioned Schema, Europe's leading independent management consultancy in the technology and media markets, to conduct a survey on broadband entertainment services [9]. Schema surveyed 5 000 households in North America and Europe to determine their attitude towards, and willingness to pay for, broadband entertainment. It remains the most comprehensive study done on the subject and is used widely in this chapter.

2.4.1 Television and video

At the time of writing, broadband TV and broadband video are two very distinct activities, though most pundits agree that they will converge during the latter part of the decade. Currently, digital broadcast TV is almost entirely limited to the TV set (with a set-top box), while on-demand or shared video is largely confined to the PC.

It is likely that broadcast TV and VOD will form part of a wider entertainment and communications portfolio that includes telephony, broadband Internet, information services and commercial services. The factors that will make it successful are, from the TV perspective, personalization and interactivity, and, from the perspective of the larger portfolio, one bill, one address for complaints and problems and a wide and exciting portfolio of services.

Alcatel/Schema defined a video service as 'a location where consumers go to search, download and view videos on their PC or TV.' The scope of the content concerned took in movies, TV programs, live events (sports, music) and private video content from peers. The research found that 29 % of Western European consumers and 36 % of American consumers describe themselves as quite likely or very likely to pay for online video services. The main drivers are likely to be personal video recorders (PVRs), which enable the time – shifting of TV programming, and new portable devices. Services are likely to be a combination of pay-per-view (or video-on-demand) and subscription.

According to Schema's research, subscribers are expected to spend \$11–\$40 for digital broadcast TV and another \$8–\$20 for additional video services (such as VOD) [9]. Home video spending will increase by \$29 a month over the next five years, according to the Yankee Group [10]. Of that, \$12 will be transferred spending (\$7 from PPV and \$5 from video rentals) and the rest will be growth – \$4 a month on subscription VOD (SVOD), \$8 on HDTV and \$15 on PVR. HDTV stands for High-definition TV and is technology that significantly increases the resolution of video signals, offering vastly improved picture quality over the current NTSC or PAL standards.

Yankee, in another report [11], caution that, while the opportunity is significant, the obstacles are great: 'Telcos and alternative broadband operators face a huge challenge in competing against the established delivery platforms. Cable or DTH television services have a strong presence in most European markets and there are limits to the willingness and ability of customers to pay for additional services.'

Both Yankee and Alcatel/Schema list a number of critical success factors. These include:

- combating video piracy;
- having access to quality, affordable content;
- having efficient and easy-to-use payment systems;
- developing appropriate and segmented content packages.

2.4.2 Games

Home gaming was a \$10 billion-plus business in mid 2003. Of that, console games accounted for about 60 % and PC games for about 40 %. Online volumes have taken off since Microsoft's Xbox Live launched broadband-based, multiplayer gaming in June 2002, followed shortly afterwards by Sony and Nintendo. The number of US households connected to one or more games services on a games console is expected to climb from 110 000 in 2000 to five million in 2005.

There are two ways of playing broadband games from home – on the PC and on a games console. In the PC category, there are also two types of multiplayer game – network-based and massively multiplayer online games (known as MMOGs.)

Network-based games use peer-to-peer technology. Users use websites to find and match up with partners, but after that, all information is passed directly between the players. Some of the most popular shooting games belong to this category.

Online games publishers typically make their revenue from retail sales of their games, without continuing revenue generation. Thus, it is a widely-held industry view that online games will not be a major generator of broadband revenues.

MMOGs, on the other hand, use a client – server architecture and all traffic passes through a network server. These are usually of the role-playing variety (MMORPGs), and thousands of players can participate simultaneously. This is where the true money is likely to be in a massively broadband world. *The Legend of Mir*, a phenomenally successful game in Asia, had over 60 million paying subscribers in mid 2003, according to its Korean developer, Wemade Entertainment [12].

Online gaming is a young people's preoccupation. In the US, 72 % of people aged 15–19 play online games, according to Schema's research, while in Europe, the percentage of gamers in the 15–24 category was 50 %. 84 % of the respondents said that they played online games more than twice a week.

Predictions of the size and revenue potential of the games market differ wildly, though they are uniformly high. Schema expects subscribers to online gaming services to spend $40–$100 per year on networked games once the service is fully launched, while Forrester Research projects subscription revenue of over $1 billion from MMOGs by 2005 – an annual growth rate of 58 %.

2.4.3 Music

Of all the online services, music is probably the best known – and the most notorious. Peer-to-peer music swapping has single-handedly reduced traditional music sales by a quarter or more, and raised the very real prospect that the CD single will be extinct within a few years. In fact, swapping is not the only form of broadband music activity. There is also the growing legitimate music sector, streaming music sites, Internet radio and a host of websites dedicated to distributing the music of aspiring rock stars.

Interestingly, while the early adopters of broadband music were primarily Internet-savvy young people under the age of 30, there has been a shift towards older consumers in recent years. In the US, music sales to people aged 45 and older have risen steadily, from about 12 % in 1991 to about 24 % in 2000 [9].

American consumers are expected to pay between $3.50 and $20 per month on broadband audio services, once they are fully launched. Jupiter Media Metrix projects an online music market of $6.2 billion in the US by 2006, while Schema forecasts revenues of $1.8 billion in Europe in the same timeframe.

2.4.4 Gambling

Gambling is a huge moneymaker, especially in the UK, and online gambling services are growing in popularity. Broadband service providers can offer significant value by providing authentication, billing services and coverage of nonbroadcast, live sports events in those countries where gambling is legal.

An online gambling service consists of locations where consumers go to buy lottery tickets, bet on sports events and play casino games. Services can range from a one-time opportunity to play bingo, to extensive sports and casino sites. Delivery methods for online gambling services include fixed Internet, interactive TV and mobile-based services.

Because gambling laws vary considerably from country to country, partnerships with local offline gambling retailers may be required to handle the complexities of each country's laws and regulations.

Although the amount of gambling varies significantly across geographic regions, approximately 30 % of Western Europeans and Americans gamble. Schema/Alcatel research shows that the early adopters of online gambling in Western Europe will be young people who already gamble, use the Internet and trust online financial transactions. Figures are not available for the US because gambling is illegal in most US states. According to Schema projections, Western European online betting revenues will grow from just under $1 billion in 2000 to $16 billion in 2005 – a compound annual growth rate (CAGR) of 78 %. From 2 %–5 % of the profits will go to network access providers. The margin share will vary, depending on local negotiations.

2.4.5 Internet

The Internet is what most people first associate with broadband. Which is not surprising, because surfing the Internet is the reason most people get broadband in the first place. But, increasingly, broadband is becoming a channel for entertainment services that are unconnected to both the Internet and the PC.

By the turn of the century, the Internet had reached the masses in most developed countries – the fastest spreading technology in history. Research by Global Reach reveals that 55 % of the world's English-speaking inhabitants, and 27 % of the speakers of European languages (excluding English) had Internet access in early 2004 [13].

The Yankee Group's European Connected Consumer Survey [2] found that online users spend more time online than watching TV or playing games. Online users believe the Internet is better value than fixed telephony, mobile telephony or TV services. Much of the value comes from the ever-changing nature of the Internet: more than half of the respondents used at least six applications every week, making the Internet a uniquely rich medium, especially when compared to TV or mobile telephony. However, the survey also tells another story: most consumers still see TV and related devices as their main source for paid content; they show a market reluctance to pay for content from the Internet.

The research indicated a strong relationship between broadband usage and audiovisual usage. Broadband users are at least twice as likely to download or stream music and video. Clearly, this is because broadband enables higher bandwidth applications and it is always on.

Another finding was the strong link between broadband usage and home networking: having a home network was the single most significant predictor of broadband subscription in the survey, suggesting the trend toward home networking will remain strong as broadband penetration continues to rise.

According to Yankee's analysis, the average online user in Europe spends about €5 per month on video and DVD rentals, and about €20 per month on videotape, DVD and CD purchases. Games add €2 per month. In addition, the average online user (including those who do not pay at all) spends €12 per month on satellite or cable TV. Hence, the total available market, from the Internet service provider's point of view, is about €40 per user per month. In other words, as things stand, the average Internet user spends slightly more on content per month than on broadband Internet access.

The conclusion is that broadband profits, aside from connectivity, will be made from content and services that are typically not supplied by the consumer's network provider. The broadband era will see ISPs expanding their service offerings considerably in order to stay in business.

2.4.6 Mobile

With the advent of new, higher capacity mobile networks in the form of 3G networks, as well as expected improvements in the technologies to provide content to mobile devices both within and outside the home, it can be expected that a lot of the entertainment services provided to wired broadband devices will also be made available to mobile devices. This is in addition to entertainment services provided to mobile devices based on their context and location.

The expected higher bandwidth that will be provided on the 3G networks will probably allow for streaming video type services in excess of 300 kbps, and for audio at 128 kbps. One example of these new technologies could be hybrid networks, such as DVB-H broadcast to mobile devices, which utilize the cellular network for providing return path connectivity, and thereby allow interactive entertainment on mobile devices.

In this type of hybrid network, it can be expected that there will be multiple services in excess of 300 kbps. The number of services will be dependent on the amount of capacity that will be allocated in this multiplex for this type of service. Another example could be high-capacity fixed wireless delivery of content to the home, thereby replacing the normal 'last mile' wired connection.

These networks must be able to provide audio and video in both live and stored formats and with sufficient capacity to allow a user to be able to use the network for entertainment on the go. These networks will also need to address the issues relating to protection of the content, as well as the QoS of the service provided.

For wireless entertainment within the home, there needs to be an even greater emphasis on the issues that are important for mobile users, with the additional requirement that the bandwidth that can be used will have to be on a par with high-bandwidth wired solutions. For the home networks, it is doubtful whether commercial video services will really take off until there is widespread availability of secure receivers supporting new codecs, such as MPEG-4 Part 10 (also known as H.264), which supports MPEG-2 quality video at about 60% of the bandwidth. That will mean standard definition video at about 1.5 Mbps and high-definition video at 5–6 Mbps. With an average of 2.5 set-top boxes (STBs) in a US home, a service provider will need to provide between 9 and 15 Mbps to support the average home. That is possible with ADSL2+.

In the home networks, the bandwidth that will be required is likely to be similar to the bandwidth capacity for existing digital video broadcasts, therefore, for standard broadcast, it is likely that at least 6 Mbps will be required per service, and higher bandwidths for high-definition content. For audio content, lower bandwidth will be required, once again probably in the region of 128 kbps for each audio stream.

2.5 Consumer demographics

'Young people are the early adopters of broadband services – the leading edge of a mass market that promises serious rewards for service and content providers. However, the youth market will not be easily won. It is a technologically savvy group, and their online expectations have grown with their online experience.' Thus begins an Alcatel white paper entitled *Broadband isn't Wasted on the Young* [14]. The premise that the young are in the vanguard of broadband adoption is not surprising. It has been borne out by several research projects and is self-obvious to anyone with more than a passing interest in broadband activities.

The age differentiator was also the major finding of the Yankee Group's European Connected Consumer Survey, conducted in July 2003. Age has a 'big impact,' Yankee found, both on online behavior and attitudes to new services. Other prominent Yankee findings were that gender is a relatively unimportant differentiator, though it is significant for certain services, and income level and household size are surprisingly unimportant.

As regards specifically broadband applications, such as video and music, the Yankee survey found that users over 55 make little use of 'rich media' applications, even though they are slightly more likely to have a broadband connection, while the under 20s are far more 'communication-intensive' than any other age segment.

Whether the broadband uptake of young people is commensurate with their buying power was the subject of several surveys in 2002 and 2003. A survey by research firm InsightExpress [15] found that students, including teenagers and young adults, drive the decision when it comes to consumer electronics purchases. For PCs, for example, the survey found that students made the decision in 52 % of the cases, with their parents deciding in 30 % of the instances and the two making a joint decision just 18 % of the time. The same trend was true as regards Internet access, cell phones, digital cameras, PDAs, MP3 players and software. In the latter case, students made the decision 76 % of the time.

There is no empirical evidence that the same proportions hold true when it comes to paying for broadband content, though it is a fair assumption that young members of the household have an inordinate influence over those decisions.

With telephony, data and video networks converging, service providers have begun offering service bundles to their customers, with discounts offered on the basis of the number of services included in the bundle. The basis of the bundle is the so-called 'triple play' – voice, data and video – though wireless connectivity and mobile are also items that will be increasingly prominent in bundles as the decade progresses.

The average revenue per user (ARPU) for a standard menu of services was $80 per month in the US and $74 per month in Europe in 2002, according to Schema [9]. This revenue, for a content menu providing basic TV, pay TV, personalized TV, music, games and broadband Internet access, is in addition to the telephony revenues generated by the subscriber.

As the market matures, Schema expects the ARPU for the same bundle of services to decline to $45–$50 per month by 2007. However, service providers will, no doubt, introduce additional services, thus keeping the ARPU somewhat stable.

The key point is that a household may have a broadband connection, but the consumers in the household are individuals and each has his/her own interests. Broadband will truly come into its own when it is no longer regarded as a pipe into a home (i.e. one product with one price), but the enabler for a wide range of entertainment options, each of which is purchased and consumed by one or more family members.

2.6 Paying for content

Broadband commerce presupposes not only that consumers will be prepared to pay for the content they receive over public networks, but that the commerce can be conducted securely. Without the security of knowing that copyright will be respected and goods paid for, neither the content rights holders nor the network operators are likely to commit themselves to broadband commerce in any meaningful way.

Unlike accessing your bank statement online, say – in which the transaction must be secure but the data, once received, belongs to you – content security makes no assumption of trust regarding the receiving party. Instead, it assumes that, while the receiver may well be legitimate, his/her *intentions* may well be illegitimate. In the Pay-TV business, therefore, the objective of content security is to prevent:

- the circumvention of user rights – i.e. to ensure that users receive only what they pay for; and
- the illegal copying and redistribution of content and/or the keys used to encrypt the content.

Systems which provide the above functionality – rights management and persistent content protection – are known as Digital Rights Management (DRM) systems. DRM is still a rather fuzzy concept and the disparities between systems using the name are wide. One reason could be that most early DRM systems attempted to provide an all-in-one solution, while the reality is that different types of content probably require different levels of security. The protection of a latest-release movie, for example, is likely to be far more stringent than that applied to a digital poem.

At its most rigorous, a DRM system is likely to encompass:

- encryption and decryption of content;
- secure distribution of keys (or control words);
- authentication of the receiving device;
- a means of ensuring payment for content consumed;
- entitlement of the user to the services that have been paid for.

One critical concept is *persistence*. Until recently, only PCs had hard drives; all other consumer devices could play or display digital content but not store it. That began to change with the introduction of personal video recorders in 2000. By the end of this decade, virtually every consumer device will have the ability to store content digitally and replay it as required.

From a security perspective, the abundance of content storage means that not only the distribution of content needs to be protected, but the storage as well. Content needs to remain encrypted on the hard disk and only be decrypted on rendering, and once rights issues have been dealt with.

Another critical concept is super-distribution (or peer-to-peer distribution) – the ability for one consumer to distribute content to one or more others, *while maintaining the original rights of the content*. That could mean anything from limiting the amount of people who can receive the content to ensuring that the receivers of the content pay for it before it is rendered.

DRM technology is still immature, and super-distribution exists more as a concept than a deployed technology. Companies vying to provide solutions range from InterTrust, with a fully software-based solution (and \$440 million after winning a patent infringement suit against Microsoft) to conditional access vendor NDS, with its silicon-based Secure Video Processor.

2.7 Security standards

International standards bodies and industry groups have produced, and are currently working on, a variety of technical standards relevant to broadband networks. Of the many issues being addressed, one of the most important deals with the protection of both the broadband services and the content carried by the services. If it is not possible to generate revenue from the services, there will be no services, and if it is not possible to protect the content, there will be no content for these services.

The following are examples of some of the organizations that are addressing these security issues.

2.7.1 DVB (www.dvb.org)

This organization deals with standards relating to the broadcast of content. It has various working groups dealing with both the definition of methods for distributing content in IP-based broadcast networks (e.g. TM-H, TM-UMTS), and the protection of content (e.g. TM-CPT).

2.7.2 3GPP (www.3gpp.org)

The scope of this organization is to produce globally applicable technical specifications and technical reports for a 3rd Generation Mobile System. As such, it has various working groups dealing with access control.

SA WG3 – Security

Produces specifications such as *3GPP TS 33.102: 3G Security; Security architecture,* and *3GPP TS 33.246: Security of Multimedia Broadcast/Multicast Service.*

SA WG4 – Codec

Produces specifications that address some aspects of content protection, such as *3GPP TS 26.234: Technical Specification Group Services and System Aspects Transparent end-to-end packet switched Streaming Service (PSS); Protocols and codecs,* and *3GPP TS 26.244: Technical Specification Group Services and System Aspects Transparent end-to-end packet switched Streaming Service (PSS); 3GPP file format (3GP).*

2.7.3 3GPP2 (www.3gpp2.org)

The scope of this organization is similar to that of the 3GPP organization, except that it is mandated to produce the relevant 3rd Generation Mobile System specifications specifically for North American and Asian interests.

2.7.4 OMA – Open Mobile Alliance (www.openmobilealliance.org)

The mission of the Open Mobile Alliance is to facilitate global user adoption of mobile data services by specifying market driven mobile service enablers that ensure service interoperability across devices, geographies, service providers, operators and networks, while allowing businesses to compete through innovation and differentiation. This organization has a working group that is most relevant to this environment: BAC – DL/DRM, which is currently developing version 2.0 of it content protection scheme.

2.7.5 TCG (www.trustedcomputinggroup.org)

Trusted Computing Group (TCG) is an industry standards body composed of computer and device manufacturers, software vendors and others with a stake in enhancing the security of the computing environment across multiple platforms and devices.

2.7.6 SVP (www.svp-cp.org)

Secure Video Processor is an industry group that is developing a standard secure method for digital content protection, providing the ability to transfer content from device to device in a manner that is secure and allows the content owner to retain control over the content.

2.7.7 WPA (www.wifialliance.com/OpenSection/protected_access.asp)

Wi-Fi Protected Access (WPA) is developing a specification of standards-based, interoperable security enhancements that are intended to strongly increase the level of data protection and access control for existing and future wireless LAN systems.

2.8 Summary

Broadband is in the process of redefining the concept of 'home entertainment.' Supplemented by home networking, broadband is causing the distinctions between home electronics and computers to disappear. By the end of the decade, the home will be as advanced technologically as any theater, video arcade or concert hall.

However, broadband's genesis in the Internet culture of free content could severely limit its commercial expansion. Prime content costs money to create, package and distribute. For broadband to reach its true commercial potential, it will need to respect copyright and generate a culture of payment. Failing that, broadband will remain simply a 'pipe,' without the content to justify its existence.

While the first few years of the broadband era were devoted primarily to fast Internet surfing, it will reach its maturity in the synthesis of broadband delivery with high-quality video. This will take the form of IP delivery of TV channels and video-on-demand to the TV set and, later, to the mobile device, as well as music, games, gambling, shopping and interactive advertising. Broadband will serve all members of the household and, like most technology-based services, its uptake will be driven by the young.

References

1. Houghton Mifflin Company (2000), *The American Heritage*® *Dictionary of the English Language*, fourth edition.
2. Yankee Group (2003), *Europe's Connected Consumers Love the Internet, but Their Passion for TV Still Burns.*
3. Juniper Research (2003), *Mobile Music and Ringtones*, December.
4. Reported by BBC News World Edition, 10 February 2004.
5. Reported by BBC News World Edition, 15 January 2004.
6. In-Stat/MDR (2004), *Picking Winners in On-Line Entertainment: Disney, Microsoft & Intel*, report # IN0401551CM.
7. Yankee Group (2003), *Killer App Arrives in the Form of Home Network Media Nodes.*
8. Knowledge Networks/Statistical Research (2002), *How Children Use Media Technology*, April 5.
9. Schema/Alcatel (2002), *Consumer Survey on Broadband Entertainment Services*, June.
10. Yankee Group (2003), *Do Consumers Have More Digital Video Services Than Dollars?*, research note, July 2003.
11. Yankee Group (2003), *Can Broadband Service Providers Make Money From Online Video Content?*
12. Company presentation at the BCD Forum in Washington DC, July 9–11 2003.
13. http://www.glreach.com/globstats/index.php3.
14. Alcatel (2003), *Broadband isn't Wasted on the Young*, Telecommunications Review, 2nd Quarter.
15. InsightExpress, July 2002.

3

Applications and Services to Meet Society-Related Needs

Csaba A. Szabó

Budapest University of Technology and Economics, Budapest, Hungary

3.1 Introduction

In this chapter, we are going to give an overview of a group of applications that are important for society and that can be greatly improved by broadband delivery to the users. This group includes applications and services falling into the broad categories of e-learning, e-health and e-government.

It is almost trivial that businesses need broadband applications, and there is also a viable demand for entertainment services, the only problem might be that broadband is far from being available to everyone, nor is everybody willing to pay for it. Society-related services are from a slightly different area, where it is more difficult to prove to the users that they are getting a better service, higher quality or new applications which would not be possible otherwise. On the other hand, since these services are often handled by public institutions, and the public, as we will see in later chapters, usually plays an important role in the development of broadband services, the first users of broadband networks built by some form of public participation are often the public institutions themselves.

It is also an important political issue to make these services available for every citizen in the region. In this way, at least the easier side of the digital divide – the access side – can be solved, which means that everybody can have access to broadband services, including those who live in areas traditionally underserved by telecom companies. The more difficult aspect of the digital divide is to bridge the gap between those who want to use services and those who do not.

There are important government and international programs that aim to promote broadband applications and services. As an example, the eEurope2005 plan of the European Union defines specific objectives and measures for promoting broadband access, on one hand, and the development of content for broadband, on the other.

Broadband Services: Business Models and Technologies for Community Networks. Edited by I. Chlamtac, A. Gumaste and C. Szabó © 2005 John Wiley & Sons, Ltd. ISBN 0-470-02248-5.

In the following three sections, we are going to address three key areas: *education, health care* and *governmental services*.

3.2 E-education applications

Modern communications technologies, in particular multimedia and networking, play a key role in making learning forms and resources available and accessible for all. E-learning, or e-education, is a broad term, encompassing the use of telecommunication services and the Internet in delivering education programs, together with educational management. It includes providing access for students and teachers to all possible sources of information and the like. The term e-learning also covers the enabling technologies and the appropriate – broadband – telecommunication infrastructure. Thus, as an example, connecting all schools in the EU by broadband connections is one of the goals of the eEurope2005 program within the e-learning area.

In this section, we are going to be more specific and will address two areas where broadband network infrastructure and services are of key importance for delivering distance education programs: 'virtual classrooms' and 'web-based learning'. Those familiar with the methodology of distance education know that the former is characterized by the term 'synchronous learning', while the latter can be synchronous or asynchronous, although these two main applications are not restricted to being methodologies only for distance learning.

3.2.1 Virtual classrooms

Virtual classroom type applications include:

- A *telelecture or telepresentation* which can be delivered by a person who is unable to be at the same physical location as his/her audience. We know how important it is, for instance, to bring a renowned speaker to a conference as a keynote speaker, or as a visiting lecturer whose lecture would fit well with, and extend, the series of course lectures within an academic program. The difficulties are often – mistakenly – considered as mere financial problems, i.e. paying travel and accommodation costs. There is a more serious obstacle, however: the invited person simply cannot afford – due to other commitments – the time to travel to the institution or company that hosts the program, in particular from overseas. Virtual classrooms using videoconferencing help overcome this obstacle.
- *An extended classroom.* It is often the case, both at universities and at conferences, that a lecture hall cannot accommodate all those wishing to attend, not to mention those unable to attend physically because they are at remote locations. In this case, the live presentation can be extended to further lecture halls either in the same campus or conference center, or at remote sites via two-way video/audio communications, thus making the participation at extension sites almost identical to the local attendance.
- Some *special training programs* have to be *delivered within a short time for many locations.* Training employees of a large, multisite company to become familiar with a new way of intra-company communication is one example. Or, in the public sector, it is often necessary to inform a number of organizational units spread over a large geographical area, such as

a region or a whole country, of a new procedure to be implemented in a specific area of public administration. Multipoint videoconferencing is a tool to deliver such programs, with virtually no limit as to the number of sites. In both examples, simultaneous use of document transmission is a key element.

3.2.2 Web-based learning

By this term we refer to all kinds of learning in which the educational materials are hypermedia documents and they can be reached and handled via web technologies. The network environment can be a corporate intranet and/or the public Internet. Web-based learning offers a number of advantages:

- It is predominantly a personal way of learning; therefore, the participant can define his/her own pace, order of modules to study, use of supplementary materials, etc. (hence the term 'self-paced learning').
- One-to-one interaction between students and teachers is an integral part of many web-based learning programs; tutoring is provided via email or even via online voice and data communications.
- When producing the content, the whole gamut of current multimedia technologies can be used to prepare very efficient learning tools.

Web-based learning programs are often delivered as live broadcasts over the Internet, often called 'webcasts' or 'webinars' (web-based seminars). In this case, the nature of participation is similar to the use of videoconferencing in a virtual classroom, however, the audiovisual impression is obviously poorer, and the feedback is limited to asking questions using a Q/A window. The questions can then be answered by the lecturer online during the lecture or, in most cases, via email afterwards. Web-based learning systems often allow for other forms of online activities, such as participation in discussion groups, working on collaborative projects and the like.

There are several driving forces toward the widespread deployment of the virtual classroom and web-based learning type applications. Some of these are:

- the demand for flexible forms of learning is increasing in various groups of the population;
- lifelong learning is not just a slogan, it is becoming a reality;
- there is often a need to overcome long distances;
- EU education institutions are open to applicants from all EU countries on a nondiscriminatory basis, therefore, distance education is playing an increasingly important role.

The main advantage is that students can proceed at their own pace, wherever the student resides and with great flexibility in time. There are some disadvantages, however, or rather conditions that need to be met for a successful training program:

- a certain level of IT literacy is required, together with the necessary equipment;
- the classic teacher–learner relationship is absent. It needs to be substituted by a combination of on-site lectures, high-quality virtual classroom presentations (that are only possible via

broadband) and web-based learning. Tutoring in small groups at remote locations is also an important part of the training methodology;
- it assumes more responsibility than the usual classroom course, and is therefore applicable only when participants are highly motivated.

3.2.3 Technology of e-education

The respective technology instruments are *videoconferencing* – or, using a more recent term, *media conferencing systems* – for virtual classrooms and *web technologies and audio/video streaming*, for web-based learning, respectively. In both cases, transmission of multimedia presentations is an integral part of the technology, and thus of the learning environment.

Videoconferencing is a relatively old technology and has been around for more than two decades, using expensive broadband satellite technologies and terrestrial dedicated connections at the beginning. The development of video compression technologies and, most importantly, the worldwide penetration of ISDN, starting in the early 1990s, paved the way for widespread usage of videoconferencing in several areas, including education.

Today, videoconferencing is a mature technology, but it still means what it meant two decades ago: transmission of video and voice, in compressed form, over digital telecommunication links, such as leased line, ISDN or over an IP network, and the presentation of voice and video in a synchronized way to the user. The distinctive feature of videoconferencing, as compared with television, is the use of narrowband connections. Even an xDSL link, commonly referred to as 'broadband' is considered to be narrowband, meaning it is low-speed relative to the bandwidth of analog television, measured in megahertz, not to mention digital television, where the data rate is of the order of tens of Mbps. Due to sophisticated compression techniques, videoconferencing provides a suitable video quality for most distance learning and telemedicine applications when using two or three ISDN lines (256 or 384 kbps), or, over a corresponding bandwidth of an IP connection.

As the newer name 'media conferencing' suggests, the video and voice communication is combined with data applications in an integrated way, exactly as we use these media at real working meetings, conferences and in classrooms.

Now let us have a closer look at the components of videoconferencing systems. Users are sitting in front of end systems or end points that traditionally fall into three categories: *conference room systems, compact systems* and *desktop systems*.

- *Conference room systems*, as the name suggests, are suitable for large groups, even for large lecture halls. They have large screens, often even two screens that offer various possibilities for viewing moving videos and other media documents, and high-quality sound systems.
- *Compact systems* 'know' almost everything that a large conference room system does, however, with some compromises regarding quality to ensure portability. A compact system is a kind of set-top box that is usually put on top of a TV set.
- The third category, *desktop systems*, are based on PC workstations. Once equipped with special purpose hardware to take care of the video/voice compression, desktop systems today are nothing more than a good multimedia PC with software for controlling the audio/video communication and collaboration between the parties, the latter being the most important application of desktop systems.

For multisite conferencing, a special piece of equipment called a *conference server*, or *bridge*, is needed, that can be either owned and operated by one of the participants, or rented as a service from a service provider. This special equipment is responsible for handling the video, audio and data content in multiparty calls. It switches between the sources using different modes. In voice-activated mode, as the name suggests, the louder speaker's site becomes the source of video as well. This corresponds to one of the real meeting situations which can work well with a small number of disciplined participants. For a larger number of sites and in more formal meetings, the chair-controlled mode is preferred when a chairperson is giving the floor to the next speaker. In a learning environment, the chair-controlled mode is used. A convenient feature is the ability to show to the viewers not only the active site, but all the sites, or a part of the sites, simultaneously on a screen divided into four, six or eight fields. This feature is called 'continuous presence' by some videoconferencing vendors.

The videoconferencing market has been stable, growing only slightly during the last couple of years. It seems that two major changes are underway which will facilitate the penetration of videoconferencing, in particular in the education field:

- the move from switched networks (from ISDN) to IP-based networks;
- the shift from appliance-type end points to multimedia workstations and systems based on them.

Note that another trend in the near future will be the penetration of videoconferencing into the mobile world, as the emerging 3G systems will provide an adequate platform for that.

Most of today's equipment also works on emerging telecom networks based on the Internet protocol. The public Internet, which is cheap and available everywhere, is usually not (yet) well-suited to videoconferencing, as it does not provide sufficient bandwidth, causes large delays and its behavior is unpredictable. Private IP networks, such as corporate/institutional intranets, on the other hand, are usually well-designed and properly managed, and thus provide good quality. Since a company owns the network, its costs are permanent and easy to forecast and control for budget management purposes. Telecom companies offer similar quality over their virtual private networks. The cost structure of the IP-VPN services is more difficult, so one has to study them carefully in order to plan properly.

Requirements for the delivery network include delay, packet loss and bandwidth. Delay should not be more than 200 ms, meaning end-to-end, including network delay. Delay variation, or jitter, should be in the range of 20 to 50 ms, while packet loss should not exceed 1 %. The required bandwidth depends on the frame rate and size, and the overhead specific to the network has to be added to the connection bandwidth. For example: a 768 kbps connection consumes close to 1 Mbps bandwidth, as about 20 % overhead has to be added to account for the network communication overhead. Note that current equipment supports data rates starting from 128 kbps – this corresponds to one ISDN BRI – to as high as 2 Mbps of connection bandwidth.

The main characteristics, interfaces and protocols of videoconferencing systems have been standardized to ensure interoperability of different systems, regardless of their manufacturer, design and geographical location. For IP-based videoconferencing, the ITU (International Telecommunication Union) standard is H.323, which is often called an 'umbrella' standard, since it includes interface and protocol standards, conversion between IP-based and ISDN systems (a so-called gateway functionality), and multipoint conferencing.

Webconferencing is a more recent development in media conferencing and is already widely accepted among users. It is the basis for web-based learning. Webconferencing is initially an enhancement of voice conferencing by multimedia content, such as PowerPoint slides. There is no need for special equipment, a web-browser in a multimedia workstation is the only thing needed, together with a reasonable Internet connection. A live video of the presenter can be added if there is enough bandwidth, but as experience shows, the added value is limited. This is actually a key difference as compared with virtual classrooms, the latter being video-centric, while webconferencing is more voice-centric.

The enabling technologies for webconferencing are *VoIP* and *media streaming*. The main ingredients for *Voice over IP systems* are as follows:

- digitization (as in digital telecommunication networks) plus compression and silence detection to ensure efficient transmission in the form of IP packets;
- protocols to support real-time transmission over IP networks, such as RTP;
- specific signaling protocols to set up and tear down voice connections.

Today, VoIP is widely used in Internet telephony and in IP-based local exchanges that are gradually replacing conventional PBXs.

Media streaming is a content distribution similar to TV broadcasting. The content is being broadcast from a central facility to many sites (the size of the viewing population can actually be very large). The communications network used for the media streaming can be the public Internet or any IP-based private network. Streaming refers to the feature that the content is being downloaded while viewing: a very important feature as, in this case, there is no need to download a lot of data before starting viewing. Media streaming is inherently a one-way communication, however, some feedback is possible in text form via the same user interface, using the Internet, or by phone. Content can be obtained from a live source or a previously recorded and stored content can be used. Figure 3.1 illustrates a streaming media system.

Streaming media content via the Internet can be delivered by unicast and multicast methods (Figure 3.2(a) and (b)). Using the simplest means, unicast, the network load is obviously higher than in multicast, the latter, however, requires this capability from all network nodes (routers).

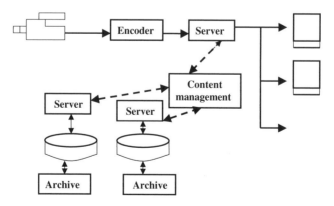

Figure 3.1   A streaming media system

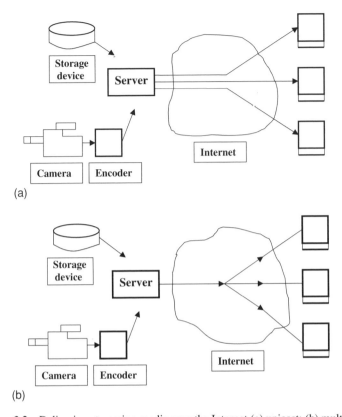

Figure 3.2   Delivering streaming media over the Internet (a) unicast; (b) multicast

Representative vendors of streaming media servers and client software systems include Cisco (IP/TV), Microsoft (Windows Media), RealSystem and QuickTime.

The user interface for a streaming-based learning application consists of several windows on the screen, the most important ones are the video window of the presenter and a second one for the textual/graphic content (e.g., a PowerPoint slide show). Video and presentation windows are properly synchronized, Just as in the case of a live presentation, when the speaker changes the slides according to the pace he/she is presenting the subject. In a streaming video application, it is also possible that the participant violates this synchronism by moving back to an earlier slide when a clarification is needed. A third window can be a questions/answers window, where the participant can write questions at any time they arise, these will be answered either online or later via email.

3.3 Telemedicine applications

3.3.1 What is telemedicine?

In this section, we will examine current and emerging telemedicine applications. Our objective is to understand the requirements these applications represent for the delivery or interconnection network, and to show that the penetration of broadband can facilitate the further development of e-health services. We will see that some advanced medical applications are impossible to introduce without having a suitable broadband communication infrastructure, while in other cases, the quality and performance of applications and services can be greatly improved by using broadband.

Let us start with some definitions and terms widely used nowadays in this context. *Telemedicine* is about a doctor–patient or doctor–doctor interaction that involves the distance element – 'tele' – and the use of suitable means of electronic communication to bridge the distance. The American Telemedicine Association formally defines telemedicine as 'The use of medical information exchanged from one site to another via electronic communication for the health and education of the patient or health care provider and for the purpose of improving patient care.'

Surprisingly enough, the first cases of telemedicine were documented almost half a century ago. An example from those early times would be providing advice and treatment to a patient on a ship from the mainland using the ship-to-shore radio link. As another example, in manned space missions some form of telemedicine was obviously used from the beginning. The most frequently cited historical evidence is the system based on a microwave link, installed between the Massachusetts General Hospital and Boston Logan Airport's walk-in clinic, in 1968. The idea was to provide immediate access to a physician when needed by a patient, without employing one permanently at the airport. Examinations included radiology, dermatology and cardiology [1].

Let us note that *e-health* is a related term, meaning the application of information and communication technologies (a commonly used European acronym is ICT) across the whole range of health care functions, from diagnostic examinations to rehabilitation, in the network of health care providers, from primary care to hospitals and within the administration of health care. The term e-health can be associated with the usual meaning of being 'Internet-based', like, for example, e-commerce. However, while the increasing role of the Internet and IP-based communications in e-health should be recognized, e-health is a broader term referring to the use of other telecommunication facilities and services as well, in fact, to the use of information technology in general. To illustrate the above, an essential element in e-health is the Electronic Health Record, which is a lifetime-long collection of patient data, including all diagnostic information collected. This is stored at an appropriate place and is used in a distributed way by the primary care nurses and physicians, as well as by specialists and hospital departments. In addition, e-health also includes the related policy, research and regulation issues.

3.3.2 Overview of telemedicine applications

Let us examine the most important applications and technologies used, and how the telemedicine applications can be improved by using broadband.

Two generic technologies build the basis for applications: *store and forward image transfer* and two-way live *audiovisual communications*. The first one is the storage of the information captured by medical imaging equipment and transmission over telecommunication networks or links. The second generic technology is what is commonly called videoconferencing. We have already discussed videoconferencing in the context of e-learning in the previous section.

In some applications, only one of the above is used. For instance, in nonemergency situations, transmission of medical images can be followed by a phone consultation between doctors during the following 24 to 48 hours. Or, a videoconferencing consultation is conducted between the patient and a nurse/general practitioner at one location and a specialist in another location. In many applications, however, a combination of the store and forward transfer and videoconferencing is used. For instance, a face-to-face consultation is often accompanied by transmission of images, or other kinds of medical data, via the same telecommunication links. Modern videoconferencing equipment incorporates the possibility of simultaneously transmitting images of diagnostic machines, measurement data and the like.

To try to understand the main applications in the really broad field of telemedicine, let us consider its *four main areas*:

- teleconsultations;
- telesurgery;
- telemonitoring;
- tele-education.

Teleconsultation is a collective name for several areas of telemedicine. *Teleradiology* is a scenario when radiological images (from an X-ray, CT, MRI or ultrasound diagnostic equipment) are being transmitted from one location to another, for the purposes of interpretation and consultations by a remote specialist. In *teledermatology*, digital images of skin conditions are captured, using special cameras, and transmitted via a videoconferencing connection, thus allowing the two sites audiovisual communication. In *telepathology*, pathology images on electromicroscopic slides, and in *tele-endoscopy*, video transmission from endoscopic equipment, are the bases for teleconsultations. An advanced form of telepathology is when a pathologist at the remote site and the advising specialist, at the central site, are engaged in a live, dynamic interaction. A *videoconferencing link* is again the core element of the set-up, combined with a remotely controlled microscopic image station. The consultant controls the microscope while conversing with the pathologist at the remote site via video and audio, and, of course, the latter can take over control as well. *Telecardiography* is about telconsultations based on the transmission of ECG waveforms, and more importantly, echocardiograms. The latter is bandwidth-intensive, as it requires the transmission of a sequence of a large number of high-resolution images. A challenging application from a communications point of view, is teleradiology for *mammography*, due to the very high spatial resolution requirements. Finally, *virtual endoscopy* is an advanced application where the 3D image, viewed normally through an optical endoscope during the examination of the patient, is constructed by a computer using a CT taken from the inner organ in question. This is then viewed by the doctor/consultant, who carries out the examination like a virtual tour. The result of the computer reconstruction can be stored at a central facility (e.g. in a PACS system; see later in this section) and made available for remote investigations/demonstrations via the telecommunication network [2].

Figure 3.3 shows a typical telemedicine setting, with its main parts being the remote (referring) site, the consulting site, the central database and the telecommunication network. At the point of care (referring site), the main functional elements of the system are: image acquisition equipment, image capturing devices, interfacing equipment and audiovisual equipment for human communication. At the central site, we have storage equipment, which can be temporary or long-term, displays and audiovisual communication equipment.

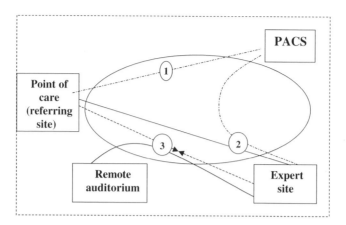

Figure 3.3 A telemedicine setting: (1) collection and storage of imaging information; (2) tele-consultation with optional use of the PACS archive; (3) distance education (arrows indicate the main direction of information flows)

3.3.3 Requirements when transmitting medical diagnostic information

The most common medical imaging equipment includes:

- Magnetic Resonance Imaging (MRI);
- Ultrasound (US), Doppler US;
- Computed Tomography (CT);
- Digitized X-Ray;
- Digital Mammography;
- Digitized Color Microscopy.

In most cases, the output of medical imaging equipment is a series of still pictures of different sizes and with different resolutions required. The pictures are produced by digitizing the conventional media, such as films, when traditional (analog) equipment is used, or when the format of digital equipment is proprietary and is not made available by the vendor. Image acquisition can also be carried out by frame grabbing from the output of the monitor. Current equipment and MRI and CT modalities produce digital images directly suitable for transmission and computer processing. The image size and resolution, and thus the resulting file size, depends on the equipment and application. Typical file sizes for different types of diagnostic equipment are shown in Table 3.1.

As we can see, the original file sizes are quite large and can result in a significant transmission time when transmitting diagnostic records over digital links of various bandwidths. A natural question arises: to what extent can the images be compressed so that they consume less storage space and transmission bandwidth? There are two basic possibilities: (a) a lossless compression, where natural redundancies are extracted from the picture without noticeable loss of quality, and (b) lossy compression, where significant reductions of file size can be achieved, at the price of a noticeable quality degradation. Compression rates are only from 2:1 to 4:1 in the first case, and 10:1 and higher in the second. Note that, in the second case, the transmission still

Table 3.1 Typical file sizes of examinations produced by most common diagnostic modalities

Modality (type of equipment)	Number of pixels	Resolution (bits)	Typical number of images per exam	File size (MBytes)
Magnetic Resonance Imaging (MRI)	256 × 256	12	60	5.9
Ultrasound (US)	512 × 512	8	20–200	5.24 – 52.4
Doppler US with color images	512 × 512	24	20–200	15.7 – 157.2
Digitized Color Microscopy	512 × 512	24	1	0.78
Computed Tomography (CT)	512 × 512	12	40	15.7
Digitized X-Rays	2048 × 2048	12	2–4	12.6 – 25.2
Digital Mammography	4000 × 4000	12	4	96

can be 'visually lossless', while being 'diagnostically lossy'. A common view in telemedicine is that for remote storage and remote diagnostics applications, only lossless compression is used, while highly compressed pictures are mainly used for teledemonstrations and distance learning. Some characteristic figures are shown in Table 3.2 for orientation.

As we can see from the table, low-speed digital transmission facilities, such as ISDN, are not suitable for transmission of most medical images, even in a highly compressed form. A 2 Mbps connection provides acceptable transfer times (a few seconds) only for compressed records. In the case where there is not adequate transmission bandwidth, sophisticated technologies are used that allow for basically lossless reproduction of pictures that are of interest to the specialist. The idea is to transmit as many pictures as possible in compressed, low-resolution format and present them to the specialist, who can choose those that are of interest. The latter are then transmitted in uncompressed form.

Teleradiology systems and the Picture Archiving and Communication System (PACS), widely implemented in hospitals and clinics, are closely related to each other and often implemented in connection with each other. PACS is essentially an intra-enterprise system that uses the enterprise's high-speed local area network for accessing the archive from workstations. Medical information is stored and transmitted according to the DICOM standard (Digital Imaging and Communications in Medicine) which, in principle, allows for interoperability between different medical systems. For a comprehensive treatment of current PACS systems, refer to [3].

The main components of a PACS system are shown in Figure 3.4. Medical imaging devices are connected, through appropriate interfaces, to the controller and archive, the latter being the heart of the PACS system. Medical staff can access the database via high-resolution display workstations. The interconnection network is a LAN. Current enterprise LANs are usually Fast Ethernet LANs, thus providing 100 Mbps data rates. The LAN is usually segmented and includes switched sections where the whole bandwidth is made available for a certain piece of equipment to be interconnected. Remote sites are connected via wide area network connections, which are currently ISDN, leased lines of various speeds, or direct optical fiber or wireless links. This is the part where broadband access can be very important, since, as we have seen, transmission of medical images via narrowband links can be prohibitively slow.

Table 3.2 Examples of transmission times of radiographic images, uncompressed and compressed, at two link data rates

Type of image	Picture size (pixels)	Dynamic range (bits per pixel)	Number of images	File size no compression (MBytes)	Transmission time, in seconds, at data rate of		File size 16:1 compression (MBytes)	Transmission time, in seconds, at data rate of	
					128 kbps	2 Mbps		128 kbps	2 Mbps
Chest X-ray	2048 × 2048	12	4	25.2	1573	100.7	1.57	98	6.3
Mammography	4096 × 5120	12	4	125.8	7864	503.3	7.86	491	31.5
CT	512 × 512	12	40	15.7	983	62	0.98	61.4	3.87
MRI	256 × 256	12	50	4.91	307	19.7	0.31	19.2	1.23

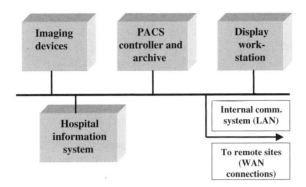

Figure 3.4   The main components of a PACS system

A relatively recent development is accessing PACS databases via the Internet. Due to sophisticated image processing, medical professionals can share medical images stored in PACS, even over low-bandwidth Internet connections. The iPACS technique is implemented in a web gateway [4]. The main idea behind iPACS is that only relatively small areas of images are processed (compressed and transmitted), so that the bandwidth need is greatly reduced. Current iPACS systems allow viewing of PACS images through a dial-up Internet connection.

In addition to still pictures, telemedicine applications require the transmission of a moving video from diagnostic equipment (or, a sufficiently long series of still pictures/frames taken at very short intervals so that at the observer's end, it can be reconstructed as a moving video). Ultrasound systems typically produce 512×512 pixel frames, with dynamic resolution of 8 bits, and for a 30 frame-per-second motion quality, a channel bandwidth of about 60 Mbps would be needed if a real-time reproduction was required. This is usually not the case; however, taking into account that a typical duration of a record is of the order of 60 seconds, a high transmission data rate is still required to allow a reasonable transmission time of the resulting long file.

3.3.4 Videoconferencing in telemedicine

Full motion, live video communication between persons or groups constitutes an essential part of today's telemedicine applications. It includes showing the environment and the patient, as required by the specific diagnostic session. This is done by using some kind of videoconferencing system, either in a point-to-point or in a multipoint setting. Multipoint videoconferencing is widely used in telemedicine for educational applications.

Figure 3.5(a) and (b) illustrate two videoconferencing systems developed for medical applications by Tandberg [5]. The powerful HCS III (Health Care System III) is implemented on a lightweight, easy-to-maneuver roll-about cart. The wide angle camera, a 20 inch medical grade monitor on a swing-arm with an adjustable height bar, and the unidirectional shotgun microphone allow for free movement around consultation rooms (Figure 3. 5(a)). Medical and nonmedical peripheral equipment, as well as a personal computer, can be added. The system supports 15 frames per second transmission when only modest data rates are available on the communication link (56 – 128 kbps) and 30 fps on 168 kbps – 3 Mbps links. Even 60 fps is possible in point-to-point configuration. It works on ISDN, using up to six BRI interfaces

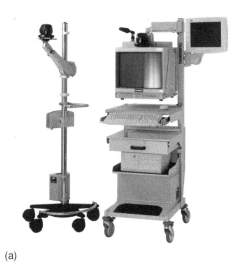

(a)

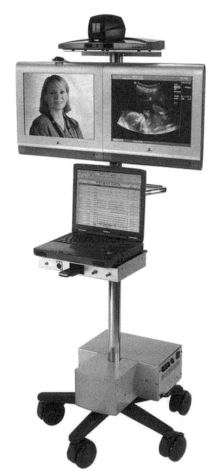

(b)

Figure 3.5 Videoconferencing systems developed for medical applications. Reproduced by permission of Tandberg. (a) Health Care System III; (b) Intern II

(768 kbps) or a PRI (2 Mbps), on a leased line with up to 2 Mbps (E1) data rate, as well as over IP through a 10/100 Mbps Ethernet interface.

The Intern II system, shown in Figure 3.5(b), is intended for applications in which a smaller, more mobile system is needed to perform remote consultations. The equipment utilizes 15 inch flat-screen LCD displays. The stand design ensures a high level of stability and maintains a small footprint, allowing ease of mobility within a medical facility.

Important, but less bandwidth-demanding, information sources are measuring instruments of *biometric* data (such as an electronic thermometer or blood pressure meter) and the output of the electronic/digital versions of some traditional diagnostic devices, such as a stethoscope or otoscope. Even for ECG telemonitoring, taking into account a 12-lead diagnostic, a typical sampling rate of 400 samples per second and a 12 bit per sample quantization, the resulting data stream can fit into a single ISDN 'B' channel.

In *telesurgery*, a surgical operation is being followed, supervised, controlled or even conducted by a remote specialist, the latter case being called *telepresence surgery*. In military environments, it is often extremely useful to have a remote specialist to advise the local surgeon; and in field uses, to accomplish the operation remotely by using robotic equipment in the operation room. In telepresence, the key technology is the robotic arm system, however, the reproduction of the visual environment at the remote site is equally important, and this latter consumes the majority of the bandwidth of the communication link, only a fraction of it is used for transmitting the robotic control and feedback information.

In general surgical practice, the most frequently used scenario is *telementoring*, which is based on an online, two-way audiovisual communication between the operation room and a remote site, using videoconferencing systems. From the remote site, specialists can follow the operation and advise on special issues or form a second opinion during the operation. This set-up is also extremely useful for conducting demonstrations for educational purposes. The videoconferencing equipment, as mentioned above, is capable of transmitting radiographic or other diagnostic pictures, as well as other diagnostic data, when needed during the operation. A typical room installation consists of a panoramic room camera showing the operation room team and the patient, a ceiling camera showing the patient, and a document camera as an input device for any textual and pictorial diagnostic information. A multipoint technique allows several remote sites to follow the operation.

The technical environment of *minimally invasive surgery* (MIS) is quite close to a telepresence operation scenario. Laparoscopy is a technique by which the surgeon conducts the operation by using manipulators and a fiber optic camera inserted into the abdomen of the patient through small incisions, and follows the whole process on a monitor. The extension to doing it from a distance is, therefore, quite straightforward. In this rapidly evolving area there is an increasing need for new specialists and, therefore, training via videoconferencing is particularly important.

Telemonitoring in home health care is another important and growing area. Remote monitoring of vital signs and patient diagnostic data by a health care professional remotely allows improved care of patients in their homes. They can be just elderly people or patients suffering from chronic diseases who live alone, or are left alone by relatives during the day. The advantages are manifold and the investment needed is quite low, as there is no need for expensive high-speed communication.

We have already mentioned education and training in several places when discussing telemedicine systems and applications. General methodology and systems of tele-education or distance education, such as videoconferencing, streaming media and webconferencing, as well as the requirements for the communication network, are discussed in the previous section of this chapter.

3.3.5 How does society benefit from telemedicine?

Telemedicine applications bring direct (measurable) and indirect benefits. Direct benefits from the increased effectiveness of medical processes include:

* savings in treatment costs, due to more efficient internal working processes of the health care organizations;
* savings in materials (such as films, paper);
* savings in travel costs, time, etc. of the medical and technical personnel;
* savings in travel costs and work time of the patients, or citizens to be screened.

Indirect benefits include:

* screenings can be made available for citizens in rural/sparsely populated areas;
* patients at local points of care can receive more efficient treatment through the possibility of remote consultations;
* in general, telemedicine contributes to the improvement of quality of life of citizens.

3.4 E-government applications

3.4.1 Overview

During the last decade, the emergence of the Internet and related technologies led to the development of e-government, or electronic government, services. E-government is a broad term, referring to e-government services themselves, the underlying applications and technology and the related nontechnical issues, such as management of e-government services and social aspects. Typical examples of e-government projects and services are as follows.

* e-government portals (at local, regional and state level, also cross-level portals);
* kiosks: public places to allow citizens to access e-government services free, or at low cost;
* digital signatures: these are a necessary ingredient of many e-government services, such as filing tax reports or using different applications;
* technology centers for communities: public places where citizens or user groups can get advice on how to use e-government services, obtain training, assistance in building networks for communities, etc.;
* e-procurement: this makes the usual public purchase procedures simpler, less costly, easy to manage and controllable by the public.

There are several nontechnical challenges:

* education of the citizen is a key to success;
* it is necessary to define optimal implementation/operation models, often with public–private partnership;

- the introduction of e-government services requires a reform of the public administration, or at least re-engineering of the processes within administrative organizations;
- users require a high level of trustworthiness – citizens expect that the government provides protection of information, reliability and availability and integrity of information systems;
- there is a danger of being technology driven.

The discussion of these extremely important social and management aspects is outside the scope of this chapter.

Technical challenges include:

- telecom infrastructure and services for public institutions;
- building and operating public access places;
- access to services in traditionally underserved areas;
- access by the method the users choose;
- access via a variety of user platforms at home and at public access places (PC, mobile, PDA, TV set).

The next section will address some of the technical challenges, primarily related to efficient, ubiquitous and multiplatform access to e-government services by citizens.

E-government applications bring direct (measurable) and indirect benefits, including:

- making transactions with government organizations faster and smoother;
- saving citizens time and expense;
- increasing the productivity of the public sector, due to more efficient intra-government operations;
- coping with unemployment;
- fostering interactions, networking among citizens, increased participation of all citizens in the government;
- improving quality of life;
- closing the digital divide.

3.4.2 Providing access to e-government services

Of the technological challenges, we are going to address the *delivery aspects* of e-government services in the rest of this chapter.

The main requirements for the telecommunication service are that it has to be (i) ubiquitous, including rural and sparsely populated areas; (ii) inexpensive for basic e-government services for the citizens; and (iii) expandable to broadband as soon as the residents want to access new, advanced services that require much higher bandwidth. Typically, traditional telecommunication companies and ISPs can satisfy none of these conditions. The service is not ubiquitous: populations in sparsely populated (rural) areas, or geographically challenged regions (mountains) remain underserved, as network deployment costs are too high compared with the expected revenues. Even the basic Internet service is often too expensive for citizens of lower income. In addition, most telcos are stuck with a single technology used for access in a particular area (copper subscriber lines, cable, wireless) and for them it is not viable economically to upgrade their access network to provide new broadband services.

For the interconnection of public institutions, broadband, reliable and cost-effective telecommunication networks are needed. Implementing government information networks by relying on facilities and services leased from telecommunication service providers is not a feasible solution, at least not in the long term, and definitely not in some areas where the telecom market has not become competitive yet, as the incumbent operator continues to preserve its monopoly. Telcos are not only an expensive solution today, but their networks cannot support emerging broadband services, or if they can, this is at prohibitively high costs.

Therefore, a number of local, regional and national governments have already decided to invest in a private network that is largely independent of telcos' facilities, is managed and maintained 'in-house' and can be upgraded or extended according to future needs, based only on the government's decision and, of course, financing possibilities. These initiatives are characterized by the terms *community networks*, or *municipal networks*, or *open access networks*, each term referring to a particular aspect of building network infrastructure based on public participation.

A multiplatform capability is a very important objective for providing e-government services, which must be accessible for all citizens, regardless of the specific access solution by which they are connected to the network. Therefore, the access part of community networks should be based on several existing and emerging technologies. xDSL, mainly ADSL (Asymmetric Digital Subscriber Line) is the most commonly used broadband access technology today. This technology utilizes the existing copper subscriber wires and provides the users with data rates up to 512 kbps in the downlink direction (i.e. from the network to the user), and up to 128 kbps in the other direction, thus it is very suitable for typical Internet usage. xDSL, however, is not a real broadband technology, many present and future applications need higher data rates, which can be provided by some sort of 'fiber to the user or to its close proximity'. FTTH (fiber-to-the-home) is a solution which provides a direct optical cable connection to the user, while in the FTTC (fiber-to-the-curb) solution, the optical cable is terminated at a single access point for a multiflat building, or for a group of small houses. The users can be connected to the distribution point via the same copper lines used by ADSL, however, the much shorter distance allows for significantly higher data rates.

The cable TV network is a similar technology, which also leverages from a combined use of optical transmission (in its star-like backbone network) and traditional cables to subscriber homes.

An emerging, and already important, access technology is WLANs (wireless local area networks) which are well-known to the general public from hotel lobbies, airport waiting areas or from exhibition and conference halls. The 'hot spots' based on this technology provide users carrying their notebooks or handheld computers with public Internet access. In a community network, a WLAN is used as a last mile solution and is adequate for several reasons: it bridges distances of several kilometers, provides 10 Mbps bandwidth using current products and will provide 54 Mbps based on new standard versions, it operates in a frequency band which is unlicensed in many countries and lastly, the equipment needed, both at the central site and in the customers' equipment, is very cheap. It is not surprising that many community networks have already implemented this technology in the United States, Australia and Europe.

3.5 Summary

In this chapter, we looked at society-related applications and services in three areas: education, health care and public administration. The importance of these applications for society was demonstrated. We saw that most of these services cannot be made accessible without broadband telecommunications, thus representing a real driving force for the development of broadband infrastructures.

References

1. Murphy, R.L.H. and Bird, K.T. (1974), Telediagnosis: A new community health resource: Observations on the feasibility of telediagnosis – based on 1000 patient transactions, *American Journal of Public Health*, **64**(2), 113–119.
2. Bartz, D., Hauth, M. and Miller, K. (2003), *Advanced Virtual Medicine: Techniques and Applications for Virtual Endoscopy and Soft-Tissue Simulation*, Tutorial at MICCAI, 2003.
3. Huang, H. K. (1999), *PACS Basic Principles and Applications*, John wiley & Sons, Inc., New York.
4. www.realtimeimage.com
5. www.tandberg.net

Part Two

Business Models

4

Key Legal and Regulatory Issues Affecting Community Broadband Projects in the United States

Sean A. Stokes and Jim Baller
Baller Herbst Law Group, Washington DC, USA

4.1 Introduction

The following provides an overview of the key legal and regulatory issues affecting municipal or public sector providers of broadband communications networks in the United States. Such providers may be subject to various federal, state and local requirements, depending on where they are located, what services they provide and how they structure their activities. In this chapter we review the legal issues that have emerged over the years as the most important to public sector providers.

Like everything else in the communications area, legal issues are constantly evolving and are becoming increasingly complex. As a result, this overview cannot cover all potentially relevant issues in depth, nor can it furnish legal advice on specific issues. Rather, the primary goal is to furnish potential public sector providers with sufficient historical background and current detail to enable them to understand the fundamentals, and to identify and locate the additional information necessary to develop comprehensive legal plans.

4.2 The benefits of public fiber-to-the-home (FTTH) systems

Access to advanced communications services is vital to economic growth, educational opportunity, affordable health care and quality of life.

In the foreseeable future, the private sector will not offer communities – outside dense population centers – sufficiently robust communications services to spur economic development, see, e.g. [1, 2].

Broadband Services: Business Models and Technologies for Community Networks. Edited by I. Chlamtac, A. Gumaste and C. Szabó © 2005 John Wiley & Sons, Ltd. ISBN 0-470-02248-5.

It is important to note here that the current generation of broadband technologies (cable and DSL) may prove woefully insufficient to carry many of the advanced applications driving future demand. Today's broadband will be tomorrow's traffic jam, and the need for speed will persist as new applications and services gobble up existing bandwidth [3].

By owning its own FTTH network, a community maximizes its ability to

- control the types, quality, reliability, timing and location of communications services deployed in the community;
- ensure that services will be available to the community at the lowest possible price;
- promote universal access and interconnectivity;
- enhance the community's economic development, educational opportunity and quality of life;
- minimize disruption to public property and maximize efficient use of public rights of way;
- improve government efficiency and communication with the public;
- enhance the local government's revenues from, and decrease its external expenditure for, communications services;
- spur incumbent providers to lower prices and improve quality of service;
- retain amounts saved within the community, where they will typically recycle four or more times.

A credible threat of municipal entry may be sufficient to cause significant changes in an incumbent's performance.

Communities that operate their own electric utilities are particularly well-suited to operating their own FTTH systems for the following reasons:

- The same technological, demographic and economic forces are at work as those existing in the early stages of the electric power industry (see [4]).
- 75 % of America's public power utilities serve communities with less than 10 000 residents. In the next 3–5 years, public power utilities may be the only viable providers of advanced communications services in many of these communities.
- Public power utilities are 'anchor tenants' that substantially reduce the financial risks of building and operating public communications networks.
- Public power utilities have more than a century of experience in providing sophisticated, technologically-complex services, billing and supporting customers of all kinds, and furnishing universal services.
- The need to survive in a decontrolled and restructured electric power industry against competitors that can bundle energy and telecommunications services.

4.3 Burdens and risks of public broadband systems

There are *financial, technological, marketing* and *political* burdens and risks faced by the public entity when planning and implementing public broadband systems.

1. *Financial:*
   - costs – construction, operations and maintenance;
   - revenues – need relatively high penetration rates;

- unavailability of suitable financing at all stages;
- restrictions on financing;
- costs of combating challenges by incumbents (legislative and judicial).

2. *Technological:*
    - changing technologies – complete obsolescence or enough competition to cut penetration below necessary levels:
        - the longer the payback period needed to reach success, the greater the likelihood of pressures from new or improved technologies;
        - include all technologies, e.g., improved performance (DSL for wireline and DOCSIS 2 and beyond for cable), Wi-Fi, fixed terrestrial and satellite wireless, Ultra Wide Broadband;
        - electric power lines – Manassas, VA pilot.
    - economies of technological scale – local vs. regional, national or international operations;
    - the need for expertise in multiple areas.

3. *Marketing:*
    - government entities generally lack communications marketing expertise;
    - national brands vs. local identity;
    - bundling of cable, local and long-distance telephone, Internet, etc.;
        - value of whole package may offset deficiencies in any individual area – e.g., poor cable service;
        - economies of scale in operations, bulk purchasing, borrowing, advertising, etc.
    - deep cost cutting by incumbent.

4. *Political:*
    - deep philosophical differences – e.g., private enterprise vs. local self-help;
    - regional differences;
    - major incumbents have vast political clout with Congress, many state legislatures.

4.4 Legal authority of public entities to provide communication services[1]

United States federal law has played a pivotal role in the communications area since 1910, when Congress enacted the Mann–Elkins Act to give the Interstate Commerce Commission authority to regulate telephone, telegraph and undersea cable providers, services and facilities. In 1934, Congress enacted the first comprehensive federal communications law, the Communications Act of 1934, which incorporated many key federal principles and requirements that continue to exist today.

In the 1934 Act, Congress established the Federal Communications Commission (FCC) and granted it broad authority to regulate 'interstate and foreign ... communication by wire and radio.' The term 'interstate' meant any service that originates in one state and terminates in another. The Act also preserved state authority over communications originating and terminating within their borders: 'Nothing in this Act shall be construed ... to give the

---

[1] The analysis presented in this section focuses primarily on publicly provided retail services. The issues surrounding wholesale services are more complex, as the exact nature and regulatory status of such services may differ from project to project and state to state.

Commission jurisdiction with respect to . . . charges, classifications, practices, services, facilities, or regulations for, or in connection with, intrastate communication service of any carrier.'[2]

The 1934 Act gave the FCC jurisdiction only over 'common carriers' – that is, entities that hold themselves out as being willing to provide services indiscriminately to all who want them, and are willing to pay the going rate. The 1934 Act required such common carriers to provide their services at just and reasonable prices; to refrain from making unjust or unreasonable discriminations; to utilize just and reasonable practices, classifications and regulations; to keep records, make reports and file tariffs in accordance with FCC requirements; to obtain FCC approval before acquiring or constructing new lines or terminating services; and to participate in FCC complaint processes. The Act did not apply similar requirements to 'private carriers' that provide communications pursuant to individually-tailored, individually-negotiated agreements.[3] This distinction between common and private carriage is an especially important one for public power utilities today.

The 1934 Act also established the precedent of dividing communications services into separate and distinct categories, each with its own history, definitions and substantive requirements. The Act separated telephone and telegraph services from radio services, regulating the former under Title II, and the latter under Title III. The FCC followed that precedent in the early 1970s, when it began to treat computer-based data and information processing services as a separate, unregulated category of services called 'enhanced' or 'information' services. Congress did so as well in 1984, when it enacted Title VI to regulate cable television services, and again in 1993, when it added a new part to Title III to deal with wireless telecommunications services. The FCC has struggled to keep these categories separate, but with the rapid convergence of services and providers today, and particularly with the explosive growth of the Internet, both Congress and the FCC are coming under increasing pressure to discard the traditional regulatory distinctions and reinvent the communications laws.

Among the most important aspects of the 1934 Act was its adherence to the prevailing economic theory that local telecommunications markets are best served by a single provider. According to this 'natural monopoly' theory, the cost of constructing, operating and maintaining a telecommunications service that offers affordable service to all addresses is so great that, as a practical matter, only a single provider can survive in a local market. As a result, regulators must preserve and protect the local monopolist by allowing it to obtain reasonable rates of return and by barring 'cherry pickers' from siphoning off lucrative business. At the same time, however, regulators must vigilantly control rates, terms and conditions of sale to prevent the telephone company from abusing its monopoly power. The theory also contemplated that the

---

[2] The physical presence of facilities within a state does not by itself confer jurisdiction upon a state public service commission – communications over those facilities must originate and terminate within the state to constitute an intrastate transmission. *Teleconnect Co. v. Bell Telephone Co. of Pennsylvania,* 10 FCC Rcd 1626 (1995) *aff'd, Southwestern Bell Telephone Co. v. FCC,* No. 95–1139 (D.C. Cir. June 27, 1997). Thus, the fact that a public power utility's telecommunications facilities typically reside wholly within a single state does not necessarily mean that communications services offered by these utilities are intrastate. Conversely, communications that originate and terminate within the same state, but which may be routed through a separate state, are, nevertheless, considered intrastate in nature.

[3] FCC, *In The Matter Of Federal-State Joint Board On Universal Service,* CC Docket No. 96–45, Report to Congress, FCC 98–67, 1998 WL 166178, ¶ 22 (rel. April 10, 1998) ('Report to Congress').

local telephone company would subsidize services to low-income or other high-cost residential consumers by charging inflated rates to businesses, long-distance providers and other users of more profitable services.[4]

Between 1934 and 1996, Congress amended the Communications Act many times to respond to significant technological, commercial, legal and other developments, and the FCC and the courts interpreted the Act on scores of occasions. As new services that defied easy classification came on the market, the FCC and the states repeatedly fought over who had jurisdiction over them. Similarly, as the lines between computer applications, data processing and telecommunications blurred, the FCC and the telecommunications industry continuously battled over whether, or to what extent, the new services should be regulated as 'telecommunications' services or left unregulated as 'enhanced' or 'information' services. Despite constant tensions, however, the Communications Act survived for decades without fundamental change.

In the early 1990s, pressure for a major overhaul of the federal communications laws began to mount. Incumbent and potential new local and long-distance telephone companies, wireless providers, cable operators, computer firms, data processors, electric utilities and entities of many other kinds were eager to enter each other's lines of business but were thwarted by the 1934 Act's cumbersome cross-ownership restrictions and burdensome requirements. By the mid 1990s, Congress was ready to try a new regulatory paradigm. Trent Lott (R-MS) voiced perhaps the most succinct expression of the new spirit in Congress when he observed, during the debates on what became the Telecommunications Act of 1996, that the primary objective of the Act was to establish a 'framework where everybody can compete everywhere in everything.' Congressional Record at S.7906 (June 7, 1995).

On February 8, 1996, the President signed the landmark Telecommunications Act of 1996 into law. The Act had been passed by vast majorities in both houses of Congress, in large part, all major stakeholders perceived that they had more to gain than to lose from the new law. In the words of the FCC, the Act 'fundamentally' changed telecommunications regulation by replacing the 'old regulatory regime' that encouraged monopolies, with a 'new regulatory regime' that requires the FCC 'to remove the outdated barriers that protect monopolies from competition and affirmatively promote efficient competition using tools forged by Congress.'[5] It is with this background that we examine the key legal issues surrounding municipal or public sector entry into broadband.

4.4.1 Federal law encourages, but does not affirmatively empower, local governments to provide communications services

4.4.1.1 Cable services

The term 'cable service' means (A) the one-way transmission to subscribers of (i) video programming, or (ii) other programming services, and (B) subscriber interaction, if any, which

---

[4] *Report to Congress* at ¶ 7.

[5] FCC, Implementation of the Local Competition Provisions in the Telecommunications Act of 1996; Interconnection between Local Exchange Carriers and Commercial Mobile Radio Service Providers; Implementation of Sections 3(n) and 332 of the Communications Act, CC Docket No. 96–98, CC Docket No. 95–185, GN Docket No. 93–252; FCC 96–325, ¶ 1 (rel. August 29, 1996).

is required for the selection or use of such video programming or other programming services; 47 U.S.C. Section 522(6).

Section 613(e) of the Communications Act, 47 U.S.C. Section 533(3), provides: 'Subject to paragraph (2), a State or franchising authority may hold any ownership interest in any cable system.' 'Any State or franchising authority shall not exercise any editorial control regarding the content of any cable service on a cable system in which such governmental entity holds ownership interest (other than programming on any channel designated for educational or governmental use), unless such control is exercised through an entity separate from the franchising authority.'

At least one court has found that Section 533(e) is 'permissive rather than empowering' – i.e., it does not furnish a federal grant of authority to provide a cable service, *Time Warner Communications Inc. v. Borough of Schuylkill Haven*, 784 F. Supp. 203, 213 (E.D. Pa. 1992); but see *Warner Cable Communications, Inc. v. City of Niceville, FL*, 911 U.S. 634, 635 (11th Cir. 1990) (Section 533(e) 'authorizes local governments to own and operate their own cable systems').

In *Marcus Cable Associates, L.L.C. v. City of Bristol*, 237 F.Supp.2d 675 (W.D.VA 2002), the district court held that, because the City of Bristol lacks explicit or implicit authority to provide a cable television service, Virginia's strict version of Dillon's Rule (see Section 4.4.2.1) requires that it be deemed to lack such authority. The court also found that a cable television service is not an essential service and is thus not a 'public utility' of the kind that the City was authorized to provide under Virginia law and the City's charter. Following enactment of legislation allowing the City to provide a cable television service effective from July 1, 2003, the City dismissed its appeal.

4.4.1.2 Telecommunications services

The term 'telecommunications service' means 'the offering of telecommunications for a fee directly to the public, or to such classes of users as to be effectively available directly to the public, regardless of the facilities used.' 47 U.S.C. Section 3(46). The term 'telecommunications,' in turn, means 'the transmission, between or among points specified by the user, of information of the user's choosing, without change in the form or content of the information as sent and received.' 47 U.S.C. Section 3(43).

Section 253(a) of the Telecommunications Act provides: 'No state or local statute or regulation or other state or local legal requirement may prohibit or have the effect of prohibiting the ability of any entity to provide any interstate or intrastate telecommunications service.'

In *Public Utility Commission of Texas*, 1997 WL 603179 (October 1, 1997), the FCC ruled that the term 'any entity' in Section 253(a) does not cover municipalities, as such. The FCC found that the Texas prohibition on municipal telecommunications activities was an exercise of state sovereignty of the 'fundamental' or 'traditional' kind ' "with which Congress does not readily interfere" absent a clear indication of intent,' and that Section 253(a) is not plain enough to satisfy the 'plain statement' standard articulated in *Gregory v. Ashcroft*, 501 U.S. 452, 461 (1991).

In *City of Abilene v. FCC*, 164 F.3d 49, 54 (D.C. Cir. 1999), the US Court of Appeals for the D.C. Circuit upheld the FCC's decision, finding that 'it was not plain to the Commission, and it is not plain to us, that Section 253(a) was meant to include municipalities in the category 'any entity.' Under Gregory, the petition for judicial review must therefore be denied.' The court did not mention Abilene's leading authority, *Salinas v. United States*, 522 U.S. 52 (1999), which was decided while Abilene was on appeal, in which a unanimous Supreme Court held that a term modified unrestrictively by 'any' must be interpreted broadly unless the statute or legislative history requires a narrowing construction.

In *Missouri Municipal League*, 2001 WL 28068 (January 12, 2001), the FCC focused on municipalities that operate their own electric utilities but still held that they are not covered by Section 253(a). Finding that Missouri law treats municipal electric utilities, and the municipalities of which they are a part, as inseparable, the FCC found that the Missouri case is legally indistinguishable from, and is therefore controlled by, Abilene.

In *Missouri Municipal League v. FCC*, 299 F.3d 949 (8th Cir. 2002), the US Court of Appeals for the Eighth Circuit reversed the FCC's Missouri decision. Disagreeing with the FCC and the D.C. Circuit, the 8th Circuit found that municipalities are commonly considered to be 'entities' and that, under Salinas and similar Supreme Court precedents, courts must assume that when Congress used the modifier 'any' in an expansive, unrestricted way, it intended that the term modified be given its broadest possible scope.

4.4.1.3 Broadband, advanced services, advanced telecommunications capability and information services

Neither the Communications Act nor the FCC has defined the term 'broadband.' The FCC defines the terms 'advanced services' and 'advanced telecommunications capability' collectively as 'services and facilities with an upstream (customer-to-provider) and downstream (provider-to-customer) transmission speed of more than 200 kbps.' In the *Matter of Inquiry Concerning the Provision of Advanced Telecommunications Capability to All Americans in a Reasonable And Timely Fashion, and Possible Steps To Accelerate Such Deployment Pursuant to Section 706 of the Telecommunications Act of 1996*, ¶ 9, CC Docket 98–146, Third Report, (rel. February 6, 2002), the FCC defines 'high speed' services as those 'with over 200 kbps capability in at least one direction.'

The Communications Act defines the term 'information service' as 'the offering of a capability for generating, acquiring, storing, transforming, processing, retrieving, utilizing, or making available information via telecommunications, and includes electronic publishing, but does not include any use of any such capability for the management, control, or operation of a telecommunications system or the management of a telecommunications service.' 47 U.S.C. Section 3(20).

Section 254(b)(3) of the Telecommunications Act expresses the national goal that 'Consumers in all regions of the Nation, including low-income consumers and those in rural, insular, and high-cost areas, should have access to telecommunications and information services, including interexchange services and advanced telecommunications and information services, that are reasonably comparable to those services provided in urban areas and that are available at rates that are reasonably comparable to rates charged for similar services in urban areas.'

Section 706(a) of the Telecommunications Act requires the FCC and the states to 'encourage the deployment on a reasonable and timely basis of advanced telecommunications capability to all Americans (including, in particular, elementary and secondary schools and classrooms) by utilizing, in a manner consistent with the public interest, convenience, and necessity, price cap regulation, regulatory forbearance, measures that promote competition in the local telecommunications market, or other regulating methods that remove barriers to infrastructure investment.'

Section 706(b) of the Telecommunications Act requires the FCC to determine annually whether 'advanced telecommunications capability is being deployed to all Americans in a reasonable and timely fashion,' and if the FCC's determination is negative, to 'take immediate action to accelerate deployment of such capability by removing barriers to infrastructure investment and by promoting competition in the telecommunications market.'

While Section 253(a) expressly applies to 'telecommunications service(s),' it prohibits 'effective' prohibitions as well as explicit prohibitions. In ¶ 22 of its *Texas Order*, the FCC stated: '[S]ection 253 expressly empowers – indeed, obligates – the Commission to remove any state or local legal mandate that 'prohibit(s) or has the effect of prohibiting' a firm from providing any interstate or intrastate telecommunications service. *We believe that this provision commands us to sweep away not only those state or local requirements that explicitly and directly bar an entity from providing any telecommunications service, but also those state or local requirements that have the practical effect of prohibiting an entity from providing service.* As to this latter category of indirect, effective prohibitions, we consider whether they materially inhibit or limit the ability of *any competitor or potential competitor* to compete in a fair and balanced legal and regulatory environment.'

Under the foregoing standard, is a state barrier on cable or Internet service an 'effective' barrier to the provision of a 'telecommunications service' if a potential provider's business plan shows that inability to provide *all* such services destroys its ability to provide *any* of them?

4.4.2 State laws affecting the authority of public entities to provide communications services

4.4.2.1 Dillon's Rule states

Under 'Dillon's Rule,'[6] the authority of a municipality is strictly construed to include only those powers that the state's constitution or legislature have expressly granted to it, or that are necessarily implied or incidental to powers expressly granted.

In some states, the rule is codified, and in others it is judge-made. Occasionally, a state has both codified and judge-made versions that are in apparent conflict (e.g., South Carolina). Where the rule exists, silence is generally construed against public authority to provide communications services.

If a government entity is in a Dillon's Rule state, one must, if possible, justify the specific communications activity in question as a reasonable extension of a power otherwise granted.[7]

4.4.2.2 'Home Rule' states

In 'Home Rule' states, 'home rule' or 'charter' cities are generally deemed to be able to exercise any powers, and perform any functions, that are not expressly denied by the state's constitution or statutes, or by the municipality's own Home Rule charter. Many states – including Iowa itself

---

[6] Dillon's Rule is named for John Dillon, the chief judge of the Iowa Supreme Court, who first articulated it in *Merriam v. Moody's Executors*, 25 Iowa 163, 170 (1868).

[7] For two examples of such analyses, see an opinion by the Attorney General of Ohio finding that vocational schools can purchase hardware and software for a system to provide Internet access to its students and then offer Internet access for a fee to other entities and individuals. See http://www.ag.state.oh.us/opinions/1999/99-007.htm; and another opinion by the Attorney General of Ohio finding that a county's authority to incur debt for a telecommunications project can be derived from its authority to take actions to promote economic development and create jobs. http://www.ag.state.oh.us/sections/opinions/2004/2004%2D005.pdf.

– have wholly or partially repudiated Dillon's Rule. In Home Rule states, local governments have a great degree of autonomy and are often able to act in both a sovereign and a proprietary capacity.

It is very important to understand exactly how a state's Home Rule provision works. For example, in some states, only certain public entities are covered. Sometimes the state rule has presumptions that apply in certain situations but not others.[8] Some states have broad Home Rule language but the courts interpret the language narrowly.[9]

If a public entity would qualify for coverage by a state's constitutional or statutory Home Rule measure, one must ensure that the entity has followed, or will follow, all appropriate procedures necessary to take advantage of the measure.

4.4.2.3 State measures

Some states have expressly granted local governments broad authority to provide communications services. See, e.g., Ala. Code Section 11–50B–3; Ariz. Rev. Stat. Section 9–511(A), 9–514(A) (with voter approval); California Const., Article XI, Section 9(a) and Cal. Pub. Utilities Code Section 10001; 54 Cal. Atty. Gen. Ops. 135 (1971) and Fla. Sat. Ch. XII, Section 166.047; O.C.G.A. Sections 46–5–163(b) and 46–5–163(17) (Georgia); Oregon Revised Statutes Section 759.020; Va. Code Section 15.2–2160 (competitive local exchange services) and Section 56–484.7:1 ('qualifying communications services').

Some states authorize local governments to provide some services but not others. Examples of partial barriers include:

1. Missouri prohibits the state's political subdivisions from providing all telecommunications services and facilities other than services to telecommunications providers (under certain circumstances), services for internal use, services for medical and educational purposes, emergency services and 'Internet-type' services. Revised Statutes of Missouri Section 392.410(7). (Declared unconstitutional in the Missouri Municipal League case.)
2. Tennessee bans municipal provision of paging and security services but allows provision of cable, two-way video, video programming, Internet and other 'like' services only upon satisfying various anticompetitive public disclosure, hearing and voting requirements that a private provider would not have to meet. Tenn. Code Ann. Section 7–52–601 et seq.
3. Washington expressly authorizes Public Utility Districts to provide wholesale telecommunications services, but not retail services. RCW Section 54.16.330.

Some states have enacted outright prohibitions on municipal telecommunications activities. Examples are:

1. Texas bars municipalities and municipal electric utilities from offering telecommunications services or facilities, directly, or indirectly, through private telecommunications providers. Texas Pub. Util. Code Section 54.202 et seq.

---

[8] Opinion of the State of Washington Attorney General, AGO 2001–3. Although municipal corporations must have explicit or implicit authority, they enjoy presumption in their favor.

[9] See, e.g., Thomas S. Smith, *No Home on the Range for Home Rule*, 31 Land & Water L. Rev. 791 (1996).

2. Arkansas prohibits municipalities from providing local exchange services. Ark. Code Section 23–17–409.
3. With certain limited exceptions, Nevada precludes cities with populations of 25 000 or more from offering any telecommunications services, as defined in the federal Telecommunications Act. Nevada Statutes Section 268.086.

Some states have enacted measures that are not explicit prohibitions but impose burdens that are difficult, if not impossible, to meet. For example,

1. Minnesota requires municipalities to obtain a 65 % super-majority vote in order to provide telecommunications services. Minn. Stat. Ann. Section 237.19.
2. South Carolina allows public entities to provide telecommunications services subject to various imputed-cost requirements. S.C. Code Section 58–9–2600.
3. Virginia now allows localities to provide local exchange and other communications services, but subjects localities to various onerous burdens. Va. Code Sections 15.2–2160, 56–265.4:4. Additional burdens have been imposed with respect to cable service offerings.

Several state legislatures recently considered (or are considering) new restrictions on public communications providers. Others considered, or are considering, measures to decontrol incumbent activities.

4.4.3 Local restrictions

Local ordinances, charters, franchises, pole agreements, bond restrictions, contracts, etc., may contain explicit or implicit barriers to entry. For example, the City of Alameda, CA, had a charter provision that precluded it from establishing any new utilities without a 2/3 vote of the electors – a practical impossibility. The city was able to, and did, eliminate the charter provision by a simple majority vote.

4.5 Involvement models and structures

Public communications projects have come in many shapes and forms; some of these involvement models can be applied to FTTH projects. Examples include:

- publicly-created, independent communications entities – e.g., Memphis Networx;
- public communications services only to government and educational users – e.g., Portland Integrated Regional Network Enterprise (OR); Milwaukee (WI) and many others;
- publicly-owned communications utilities providing only wholesale services to retail providers – e.g., NOANet (WA and OR);
- publicly-owned communications utilities providing retail services to the public – e.g., Bristol (VA) (FTTH); Kuztown (PA) (FTTH); Tacoma (WA); Ashland (OR); Cedar Falls (IA); Glasgow (KY) and scores of others;
- non-profit entities – e.g., Georgia Public Web;
- strategic partnerships with the private sector – e.g., Hawarden (IA) (with LongLines and NIPSCO); LaGrange (GA) (with Charter); Shawnee (KS) (Shawnee Municipal Authority, Com Solutions and Systems Inc., and the Oklahoma Municipal Services Corporation);
- lease of municipal facilities to private sector providers – e.g., Anaheim (CA);

- construction and sale of municipal facilities to private sector providers – e.g., Lynchburg (VA) which sold a $3.5 million 42-mile fiber optic network to CFW Communications (now nTtelos) for $1 and, in return, received (a) The 30-year irrevocable right to use all of the fibers it had previously been using; (b) eight fibers on all new routes in the city; (c) CFW's agreement to offer a broadband service to 95 % of addresses in the city within four years; (d) the best telephone rates in Virginia for ten years; (e) hundreds of thousands of dollars worth of technical assistance and equipment discounts, (f) various other benefits;
- regional entities – e.g., Oregon Central Coast Economic Development Alliance; UTOPIA project in Utah; proposed Tri-Cities project in Illinois (lost referendum); eCorridor Project in Virginia; SweetNet County, WY; Rural MD. (NB: see *GTE Northwest, Inc. v. Oregon Public Utility Comm'n*, 39 P.3d 201 (Or.App. 2002), on a county's authority to provide extraterritorial telecommunications services);
- demand aggregation – e.g., Chicago CityNet (IL); BerkshireConnect (MA); Stillwater (OK).

In structuring involvement models and using facilities developed for other purposes in the past, it is particularly important to consider tax implications (e.g., public–private partnerships are likely to qualify as 'private use' for federal tax purposes, thus removing potential tax advantages).

4.6 Other considerations

In addition to the legal and regulatory aspects discussed in the preceeding sections, there are other circumstances that have to be taken into account:

1. *Predatory pricing and other anticompetitive conduct by incumbents* – see, e.g., Scottsboro (AL) Power Board filings and FCC response at http://www.baller.com/library-comments.html; see also NATOA's testimony to Senate Judiciary Committee, attaching extensive Baller Herbst analysis of anticompetitive practices of cable incumbents, http://judiciary. senate.gov/hearing.cfm?id=1041.
2. *Access to essential facilities and programming* – see, e.g., APPA's comments to the FCC on exclusive contracts for programming, http://www. baller.com/library-comments.html; see also *In The Matter of Implementation of The Cable Television Consumer Protection and Competition Act of 1992, Development of Competition and Diversity in Video Programming Distribution*, Section 628(c)(5) of the Communications Act Sunset of Exclusive Contract Prohibition, Report and Order, at ¶ 7, 17 FCC Rcd. 12,124, 2002 WL 1396090 (rel. June 28, 2002)
3. *Access to customers, particularly in multi-user settings* – see *In the Matter of Telecommunications Services Inside Wiring . . .* , CS Docket No. 95–184, MM Docket No. 92–260, First Order on Reconsideration and Second Report and Order at ¶ 71 (rel. January 29, 2003), at http://hraunfoss.fcc.gov/edocs_public/attachmatch/FCC-03-9A1.doc.
4. *'Level playing field' issues* – see, e.g., state statutes, cases and analysis of these issues in City of Louisville briefs and court decisions at http://www.baller.com/library-comments.html.
5. *Incumbents may attempt to resurrect arguments they lost in early legal challenges to municipal systems:*
   – Inside wiring issues – e.g., Glasgow, KY, dealt successfully with such issues in litigation in the late 1980s and early 1990s, and then Congress codified protections in Cable Act

Amendments of 1992. The FCC recently stated that the inside wiring rules apply to all multichannel video programming distributors (MVPDs) in the same manner. In essence, the order maintains the status quo, reaffirming the contract-oriented rules governing home run wiring, inside wiring and MDUs. *First Order On Reconsideration And Second Report and Order, Cable Inside Wiring*, CS Docket No. 95–184; MM Docket No. 92–260, (rel. January 29, 2003), at http://hraunfoss.fcc.gov/edocs_public/attachmatch/FCC-03-9A1.doc.

– Antitrust issues – e.g., City of Paragould, AR, dealt successfully with such issues in *Paragould Cablevision, Inc. v. City of Paragould*, 930 F.2d 1310 (8th Cir. 1991).

– 'Public purpose' issues – e.g., City of Morganton, NC, dealt with such issues successfully in *Madison Cablevision, Inc. v. City of Morganton*, NC, 325 N.C. 634, 386 S.E.2d 200 (1989).

4.7 Federal regulatory issues

4.7.1 Key definitions

The term '*telecommunications carrier*' means 'any provider of telecommunications services, except that such term does not include aggregators of telecommunications services (as defined in Section 226). A telecommunications carrier shall be treated as a common carrier under this Act only to the extent that it is engaged in providing telecommunications services, except that the Commission shall determine whether the provision of fixed and mobile satellite service shall be treated as common carriage,' Act, Section 3(44). 'Common carrier services may be offered on a retail or a wholesale basis because common carrier status turns not on *who* the carrier serves, but on *how* the carrier serves its customers, i.e., indifferently and to all potential users.' *Review of the Section 251 Unbundling Obligations of Incumbent Local Exchange Carriers . . .*, Report and Order and Order on Remand and Further Notice of Proposed Rulemaking, FCC 03–36 (released August 21, 2003), at ¶ 153.

The term '*telecommunications service*' means 'the offering of telecommunications for a fee directly to the public, or to such classes of users as to be effectively available directly to the public, regardless of the facilities used.' Act, Section 3(46).

The term '*telecommunications*' means 'the transmission, between or among points specified by the user, of information of the user's choosing, without change in the form or content of the information as sent and received.' Act, Section 3(43).

The term '*information service*' means 'the offering of a capability for generating, acquiring, storing, transforming, processing, retrieving, utilizing, or making available information via telecommunications, and includes electronic publishing, but does not include any use of any such capability for the management, control, or operation of a telecommunications system or the management of a telecommunications service.' Act, Section 3(20).

The term '*cable service*' means '(A) the one-way transmission to subscribers of (i) video programming, or (ii) other programming service, and (B) subscriber interaction, if any, which is required for the selection or use of such video programming or other programming service.' Act, Section 602(6).

The term '*commercial mobile service*' means 'any mobile service . . . that is provided for profit and makes interconnected service available (A) to the public or (B) to such classes of eligible users as to be effectively available to a substantial portion of the public, as specified by regulation by the Commission.' Act, Section 332(d).

4.7.2 Implications of key definitions

'A provider of a 'telecommunications service' must comply with common carrier requirements to the extent that it is engaged in providing such services,' Act, Section 3(44). Title II of the Communications Act spells out numerous additional duties of 'common carriers,' including compliance with rules governing equal access and pricing, tariffing, record keeping, reporting, participating in the Commission's complaint processes, performing studies prescribed by the Commission, etc. The Commission has relaxed some tariffing requirements on nondominant carriers.

Under Section 251(a) of the Act, each 'telecommunications carrier' has a general duty '(1) to interconnect directly or indirectly with the facilities and equipment of other telecommunications carriers; and (2) not to install network features, functions, or capabilities that do not comply with the guidelines and standards established pursuant to Section 255 or 256,' which apply to access by handicapped and disadvantaged persons (NB: the Commission has said that a telecommunications carrier can satisfy part (1) simply by interconnecting with the public switched network; to date, the Commission has not defined the duties referred to in part (2)).

Under Section 254 of the Act, and several implementing orders and decisions, all 'telecommunications carriers' and all 'other providers of interstate telecommunications' must contribute to the federal Universal Service Fund. Contributions are based on a contribution factor announced by the Commission each quarter.

Providers of 'telecommunications services' must also fulfill various privacy requirements under Section 222.

The Act also affords 'telecommunications carrier(s)' certain benefits, including the right to interconnect with the facilities of incumbent local exchange carriers on a just, reasonable and nondiscriminatory basis; the right to participate in negotiations and/or arbitrations framed by the Commission's interconnection rules and the right to receive reimbursement for furnishing services covered by the universal service program.

Providers of 'cable services' are regulated under the federal cable provisions of Title VI of the Communications Act, and are required to obtain a franchise at either the state or local level, depending on state law. An exception to the franchise requirement is that municipally-owned cable systems are not required to obtain a franchise under the federal Cable Act, as amended. (NB: as a practical matter, however, many public cable systems subject themselves to obligations that are identical, or substantially similar, to those imposed on private cable companies).

Providers of 'information services' are not subject to federal regulation or, in most states, state regulation.

Up until 2002, the FCC had declined to take a definitive position as to the regulatory classification of cable modem services. This uncertainty helped to spawn conflicting federal court opinions on the proper classification of the service. The 9th Circuit in *City of Portland, OR v. AT&T Corp.*, 45 F.Supp.2d 1146 (W.D. Or. 1999), rev'd, 216 F.3d 871 (9th Cir. 2000), concluded that a cable modem service is a type of telecommunications service. In contrast, the 11th Circuit, in *Gulf Power v. FCC*, 208 F.3d 1263 (11th Cir. 2000), held that a cable modem service is neither a 'cable service' nor a 'telecommunications service,' but an 'information service.' The Gulf Power decision was overturned by the Supreme Court on other grounds. *National Cable & Telecommunications Association, Inc. v. Gulf Power Co.*, 534 U.S. 327 (2002).

In March 2002, the FCC released a declaratory ruling in which it found that a cable modem service is an 'interstate information service' and thus not a 'cable service,' *In the Matter of Inquiry Concerning High-Speed Access to the Internet Over Cable and Other Facilities Internet Over Cable Declaratory Ruling* ... , GN Docket No. 00–185, CS Docket No. 02–52, (rel. March 15, 2002). The FCC's decision has been appealed to the 9th Court of Appeals. The elimination of cable modem services from the scope of the 'cable service' definition has mixed regulatory, financial and political implications for municipal utilities. It should allow for greater regulatory freedom, but it also eliminates cable modem revenue from the calculation of cable franchise fees payable to local franchising authorities. In *Brand X Internet Services v. FCC*, 345 F.3d 1120 (9th Cir. 2003) (petitions for reconsideration en banc denied, petition for cert expected), the 9th Circuit overturned the FCC's declaratory ruling, holding that the court was bound by the Portland panel's decision that an Internet access service is both an 'information service' and a 'telecommunications service.' The FCC has announced that it will seek further review by the Supreme Court.

In a proceeding related to its cable modem proceeding, the FCC issued a *Notice of Proposed Rulemaking* to develop a legal and policy framework under the Communications Act, as amended, for access to the Internet provided over domestic wireline facilities, *in the Matter of Appropriate Framework for Broadband Access to the Internet over Wireline Facilities* ... , CC Docket Nos. 02–33, 95–20, 98–10, *Notice of Proposed Rulemaking* (rel. February 15, 2002). In the NPRM, the FCC proposed to rule that an Internet access service over wireline facilities is also an *interstate* 'information service,' rather than a 'cable service' or a 'telecommunications service.'

After examining the statutory definitions of 'telecommunications,' 'telecommunications service,' and 'information service,' the FCC tentatively concluded in the *Wireline NPRM* that providers of wireline broadband Internet access services should properly be classified as 'information service' providers under the Act, rather than as providers of 'telecommunications services.'

In its *Wireline NPRM*, the FCC sought comment on the potential consequences of its tentative legal interpretation. Among the areas of concern were whether the FCC's proposed action would significantly impair the development of competition, as envisioned in the Telecommunications Act, and whether the ultimate effect of moving an increasing number of services out of the definitions of 'telecommunications services' or 'telecommunications' would be to bankrupt the federal Universal Service Program. The FCC's action also implicated access safeguards, interconnection, security, consumer protection and a host of other important issues.

The FCC is not likely to complete this rulemaking until the controversy surrounding its cable modem declaratory ruling is finally resolved.

The regulatory status of Voice over Internet Protocol (VoIP) has created considerable controversy.

1. In *Vonage Holdings Corp. v. Minn. PUC*, 290 F. Supp. 2d 993 (D. Minn. 2003), the district court ruled that Vonage's version of VoIP is an 'information service,' and permanently enjoined the State of Minnesota from seeking to regulate that service as a 'telecommunications service.'

2. On February 12, 2004, the FCC ruled in *Petition for Declaratory Ruling that Pulver.Com's Free World Dialup is neither Telecommunications nor a Telecommunications Service*, WC Docket No. 03–45 (rel. Feb. 5, 2003), that Pulver's form of VoIP, a free, purely peer-to-peer

service that operates entirely in the IP space and never touches the public switched telephone network, – is an 'information service,' rather than a 'telecommunications service.'
3. On February 12, 2004, the FCC also initiated a major rulemaking to determine how VoIP should be classified and regulated, Docket No. WC 04–36, Notice of Proposed Rulemaking (FCC 04-28) (not yet issued as of March 10, 2004).
4. Numerous states are also considering how VoIP should be classified and regulated.

Providers of commercial mobile services – i.e., commercial wireless services – are subject to minimal regulation. Section 332 of the Communications Act imposes minimal federal regulation. Section 332 preempts state and local regulation of the rates of, or entry into, commercial mobile services. State and local governments can only regulate certain 'other terms and conditions,' such as customer billing information and practices; billing disputes and other consumer protection matters; facility siting issues (e.g., zoning); transfers of control; the bundling of services and equipment; and the requirement that carriers make capacity available on a wholesale basis.

4.7.3 Access to incumbent network elements under Section 251(c)(3)

The Telecommunications Act of 1996 requires incumbent local exchange carriers to make elements of their networks available on an unbundled basis to competitive telecommunications carriers at cost-based rates, if lack of access to a given nonproprietary element would impair (from a barrier-to-entry perspective) the competitive carrier's ability to provide the telecommunications services it seeks to offer, 47 U.S.C. Sections 251(c)(3), 251(d)(2), see, *Review of the Section 251 Unbundling Obligations of Incumbent Local Exchange Carriers...*, Report and Order and Order on Remand and Further Notice of Proposed Rulemaking, FCC 03–36 (released August 21, 2003), at ¶ 84 (*'Triennial Review'*). If the incumbent's network element is deemed proprietary, the competitive carrier must meet a higher standard than 'impair,' and must show that the element is 'necessary,' Section 251(d)(2).

In March 2004, the US Court of Appeals for the D.C. Circuit reversed portions of the FCC's Order in *US Telephone Association v. FCC*, particularly with respect to continuing state jurisdiction to determine elements of the UNE platform in the future. The agency has not yet announced whether it will appeal the decision. Under the FCC's Order to obtain access to a UNE, a requesting carrier must use the UNE to provide at least some services on a common, rather than private, carriage basis. Id., ¶ 150. The FCC has concluded that 'network element' refers to 'an element of the incumbent LEC's network that is capable of being used to provide a telecommunications service.' Id., ¶ 56.

Unbundling requirements for certain individual network elements include:

- *Mass-market loops*. Incumbent LECs must offer unbundled access to stand-alone copper loops and subloops for provision of narrowband and broadband services. Incumbent LECs do not have to provide unbundled access to the high-frequency portion of their loops (the portion that enables DSL services).
- *Fiber loops*. Only in fiber loop overbuild situations, where the incumbent LEC elects to retire existing copper loops, must the incumbent LEC offer unbundled access to those fiber loops, and then only for narrowband services. Incumbent LECs do not need to offer unbundled

access to newly deployed fiber loops, or to packet switching features of their hybrid loops. *Triennial Review,* ¶¶ 247–285.

- *Enterprise market loops.* Following the August, 2003 *Triennial Review* Order, incumbent LECs are no longer required to unbundle OCn loops. Incumbent LECs must offer unbundled access to dark fiber loops, DS3 and DS1 loops, except where states (pursuant to Commission authority to conduct a more granular review) have found no impairment, when measuring the availability of feasible alternatives. Id., ¶¶ 298–331.
- *Subloops.* Incumbent LECs must offer unbundled access to subloops necessary to access wiring at, or near, a multiunit customer premises, at any technically feasible point, including a pole or pedestal, a network interface device (NID) or a feeder distribution interface. A requesting carrier accessing a subloop on the incumbent LEC's network side of the NID obtains the NID functionality as part of the subloop. Incumbent LECs must offer unbundled access to the NID on a stand-alone basis to requesting carriers. Id., ¶¶ 343–356.
- *Packet switching.* Incumbent LECs are not required to unbundle packet switching, including routers and Digital Subscriber Line Access Multiplexers (DSLAMs), as a stand-alone network element. Id., ¶¶ 355–357.

4.7.4 Pole attachments

4.7.4.1 Key definitions

As amended by Section 703 of the Telecommunications Act, Section 224 of the Communications Act of 1934 imposes on every 'utility' a broad range of duties concerning pole attachments.

Section 224(a)(1) defines 'utility' as 'any person who is a local exchange carrier or an electric, gas, water, steam, or other public utility, and who owns or controls poles, ducts, conduits, or rights-of-way (hereafter collectively 'pole attachments') used, in whole or in part, for any wire communications. *Such term does not include any railroad, any person owned by the Federal Government or any State*' (emphasis added).

Section 224(a)(3) defines the term 'State' as 'any State, territory, or possession of the United States, the District of Columbia, or any political subdivision, agency or instrumentality thereof.'

Thus, public power utilities are exempt from federal regulation of their poles, attachments, ducts, conduits and rights of way.

4.7.4.2 Major implications of the 'municipal exemption'

Public power utilities do not have to apply the specific federal access, rate or procedural requirements, but should nevertheless pay close attention to them because:

- some states have incorporated federal requirements;
- federal requirements are often viewed as benchmarks;
- Congress may eliminate exemption;
- allowable rates under federal rules may be higher than current charges;
- public power entities that provide cable or telecom services have attachment rights under federal law.

Government entities are, however, subject to Section 253's ban on barriers to entry and nondiscrimination requirements:

- unreasonable rates, terms or conditions, or substantial delay in processing applications, can arguably be barriers to entry;
- legislative history and recent cases hold that 'nondiscriminatory' does not necessarily mean 'equal.'

In conclusion, public power utilities have substantial flexibility, but cannot discriminate unreasonably.

4.7.5 Universal service

Section 254 of the Communications Act creates a new Universal Service Program that is intended to ensure that all Americans, including those in rural, insular and high-cost areas, have access to certain basic telecommunications services now, and to more advanced services in the future. The program will also subsidize a portion of the costs of furnishing access to certain additional services to schools, libraries and nonprofit rural health care facilities.

The basic concept underlying the Universal Service Program, as interpreted by the Commission, is that all 'telecommunications carriers' and 'other providers of interstate telecommunications' should underwrite the above-average costs of those telecommunications carriers that are willing, or are compelled, to provide the services covered by the Universal Service Program.

The Act established a Federal–State Joint Board that studied universal service reform and made recommendations to the Commission on November 8, 1996. The Commission adopted final regulations on May 8, 1997. The Commission has subsequently amended some of these rules in several Orders on reconsideration.

The Commission adopted most of the recommendations that the Federal–State Joint Board had made on November 8, 1996, but it justified its decisions on many issues on different and arguably more defensible grounds.

The major features of the new program include the following. Universal service support will be available initially for the following basic services:

- voice-grade access to the public switched network, including, at a minimum, some usage;
- dual-tone multifrequency signaling or its equivalent;
- single-party service;
- access to emergency services, including access to 911, where available;
- access to operator services;
- access to interexchange services; and
- access to directory assistance.

Any telecommunications carrier, regardless of the technology that it uses, is eligible to receive universal service support if it is a common carrier and offers, throughout a designated service area, all of the services supported by the Universal Service Program.

The federal Universal Service Program has four major components. The *High-Cost* component furnishes subsidies of approximately $2 billion annually to providers of certain 'core telephone services now, and possibly of more advanced services in the future, to persons living in rural, insular and high-cost areas. The *Low-Income* component provides subsidies of approximately $500 million annually to defray a portion of the installation charges and telephone bills of low-income persons, wherever they may be located, and to ensure that such individuals have affordable access to services similar to those covered in the High-Cost program. The *Schools and Libraries* component provides up to $2.5 billion annually to help schools and libraries obtain whatever telecommunications services they desire, as well as internal connections and maintenance of telecommunications networks. The *Rural Health Facilities* component provides subsidies of up to $400 million to help rural health care providers obtain telecommunications services at rates comparable to those in larger markets.

The Act mandates that all providers of interstate 'telecommunications services' contribute support payments to a Universal Service Fund. Utilizing its discretionary authority under the Act, the FCC also requires entities that provide 'interstate telecommunications' for a fee on a non-common carrier basis to contribute to the Universal Service Program. This requirement does not include entities such as utilities, that purely operate networks to meet their internal needs, and which are not made available to third parties for a fee. Nor does the requirement extend to private networks that are utilized to provide services to public safety or governmental entities.

All providers of interstate telecommunications services, and other providers, are required to make contributions to the fund and complete a universal service Worksheet on a biannual basis. Support payments are based on revenues generated from end users. The FCC therefore does not require wholesale carriers to contribute to the universal support mechanisms, provided that the carrier who utilizes the wholesale capacity to offer retail services makes such contributions itself. Thus, a utility that provides wholesale telecommunications capacity under a carrier's carrier arrangement will not be subject to a universal service contribution requirement, because this does not create 'end user' revenues for the utility. Nevertheless, all carriers, including carriers' carriers, are required to complete a Worksheet. The FCC has adopted a new unified 'Telecommunications Reporting Worksheet' that is to be utilized for the biannual reporting requirement.

Telecommunications providers whose estimated interstate contributions to universal service support mechanisms would be *de minimis*, are not required to contribute funds or file Worksheets (under certain conditions). The FCC has defined *de minimis* as a contribution that would be less than $10 000. Providers whose contribution is *de minimis* are nevertheless required to retain a copy of the Worksheet for three years as documentation of their exemption.

The FCC's proposed treatment of wireline Internet access services as information services, rather than telecommunications services, could potentially have an adverse impact on the Universal Service Fund by removing a significant source of revenue contributions.

4.7.6 Other important federal provisions

Additional federal provisions include:

- the cable provisions of Title VI of the Communications Act;
- Section 332 of the Communications Act, which sets forth federal requirements on providers of wireless services;
- Section 103 of the Communications Act, which allows registered public utility holding companies that would otherwise be subject to the core-business restrictions in the Public Utility Holding Company Act of 1935 to provide telecommunications services, information services and other communications services;
- Section 401 of the Communications Act, which requires the FCC to forbear from applying any regulation or any provision of the Act to a telecommunications carrier or telecommunications service, or class of telecommunications carriers or telecommunications services, in any or some of its, or their, geographic territory if the Commission determines that enforcement

of such a requirement is not necessary to ensure just, reasonable and nondiscriminatory conduct, or to protect consumers, and that forbearance is in the public interest;

• federal copyright, antitrust, tax and other laws of general applicability.

4.8 State regulatory issues

4.8.1 Certification

The discussion in this section is limited to state regulatory issues applicable to telecommunications services. Cable services are generally regulated at the local level rather than at the state level. Internet access services are typically not regulated at all, except when provided by dominant incumbent local exchange carriers under certain circumstances. Also, we do not discuss state municipal laws and the vast number of other state legal issues that apply generally to service providers.

States generally regulate, or at least require some form of filing, for almost all intrastate telecommunications service activities – e.g., at least nominal regulation of facilities-based providers of intrastate services. States typically assert this jurisdiction even if only a small amount of the services provided are intrastate.

Recent statutes authorizing municipal entry into telecommunications typically require initial approval by the state public service commission via a certificate of public convenience or comparable authorization (e.g., Iowa), and sometimes an ongoing role in overseeing compliance with statutory conditions of entry (e.g., Virginia). Applicants must typically demonstrate that they have the legal, technical, financial and managerial qualifications to provide the proposed services. States vary with respect to the degree of scrutiny involved in granting a certificate.

4.8.2 Tariffs

Most states require carriers to file tariffs with the PSC/PUC that set forth a description of the type of services to be offered, the prices of services and other applicable terms or conditions. Typically, states require that an initial tariff be filed with the certification application. In such cases, the tariff is reviewed along with the application, and the two are granted together. States are increasingly adopting a more streamlined approach to the tariff filing process for nondominant providers.

4.8.3 Annual reports

Most state PSC/PUCs require some kind of annual or quarterly report on the status of the regulated entity and the breakdown of gross revenues from intrastate services. In addition, many state reporting requirements include a compilation and summary of the disposition of customer complaints that have been filed against the carrier with the PSC/PUC.

4.8.4 Universal service and other contributions

Many states have their own universal service programs. Similarly, carriers are often required to contribute to state 911 and E-911 funds, which are utilized to support the development and maintenance of emergency call databases and systems capabilities.

4.8.5 Regulatory fees

Many states impose an annual fee on regulated intrastate carriers to recover state costs of administration. These state fees may or may not be imposed on governmental entities that act as providers of communications services. Increasingly, incumbent providers have been sponsoring state legislation requiring government entities to pay the same fees and taxes as private operators.

4.8.6 Interconnection agreements

Public power utilities that seek to provide competitive local exchange services will need interconnection agreements, and possibly collocation with the incumbent local exchange carriers. These agreements establish the terms under which a competitive entrant may interconnect and collocate its facilities, purchase unbundled network elements or resell the incumbent's services. Subject to the FCC's broad oversight and parameters, state PSC/PUCs oversee negotiations on interconnection agreements, and administer and enforce them once they are finalized. Entrants can either 'opt in' – i.e., use the terms of an existing agreement that the incumbent has entered into with another competitive local exchange carrier within the state – or negotiate a new agreement. If the parties cannot agree on the terms of such an agreement, then it will be submitted to the state public service commission to 'arbitrate' the dispute and render a decision within nine months from the commencement of negotiations.

4.9 Summary

In this chapter, we gave an overview of the rather complex legal and regulatory environment affecting public sector providers of broadband communication services. We showed that, at federal level, the communication law encourages, but does not affirmatively empower, local governments to provide communication services. At state level, local government in many cases has as great degree of autonomy. It is also the case that local governments are authorized to provide some services, but are not allowed to provide others.

On the regulatory side, we discussed different aspects of competition, the question of providing universal service and regulations regarding 'advanced' services based on new technologies, such as voice over IP.

The objective of this chapter was to furnish those involved in planning public participation in providing municipal, or regional, broadband services with the legal and regulatory background, thus enabling them to identify and articulate specific questions to be answered.

References

1. Baller, J. and Stokes, S. (2001), *The Case for Municipal Broadband Networks: Stronger Than Ever* (Fall 2001), available at http://www.baller.com/library-articles.html.
2. Broadband Reports Interview, available online http://www.broadbandreports.com/shownews/28553.
3. Office of Technology Policy (2002), *Understanding Broadband Demand: A Review of Critical Issues*, US Department of Commerce, September 6, 2002.
4. Baller, J. and Stokes, S. (1999), *The Public Sector's Authority to Engage in Telecommunications Activities*, available at http://www.baller.com/library-articles.html.

5

European Telecommunication Law and Community Networks

Giovanni Pascuzzi and Andrea Rossato
University of Trento, Trento, Italy

5.1 Introduction

The aim of this chapter is to provide a brief overview of the general principles governing the European telecommunications regulatory framework, with special reference to the deployment of local community networks.

For the scope of this chapter we will use a very broad definition of community networks: local telecommunications services provided for the specific needs of a geographically defined community. We will analyze the legal requirements for providing these services in the liberalized European market.

The European telecommunications regulatory framework is a complex body of law, which has evolved rapidly over the last fifteen years. The adopted approach, deemed appropriate within our limited space, is to make the legislation speak for itself, in order to provide direct access to it. Critical discussion of the rules is beyond the scope of this chapter, although some basic bibliographic references will offer an opportunity for deeper analysis of some of the issues raised in these pages.

After a short historical background, the New Regulatory Framework (NRF) will be outlined in some detail. The regulatory framework consists of a set of directives. European directives must be implemented by Member States through the enactment of appropriate legislation. We will offer a very short analysis of the Italian implementation.

Finally, we will provide a case study of community network services, provided within the legal framework, resulting from the process of liberalization and harmonization, which had a profound effect on the structure of the European electronic communications market.

Broadband Services: Business Models and Technologies for Community Networks. Edited by I. Chlamtac, A. Gumaste and C. Szabó © 2005 John Wiley & Sons, Ltd. ISBN 0-470-02248-5.

5.2 Historical background

In 1986, the European Commission published a Green Paper entitled *Towards a Dynamic European Economy, Green Paper on the development of the common market for telecommunications services and equipment* [1]. The telecommunications market in Europe, at that time strongly controlled by governmental bodies through monopolies and exclusive or special rights, was expected to grow exponentially year on year. The number of communications services was deemed to increase at an accelerated rate. Moreover, the European Economic Community was rapidly approaching the 1992 target for the completion of a common European-wide market for goods and services. Creating an open and competitive environment in the telecommunications market was seen as the prerequisite to boosting innovation and consumer opportunities on the one hand, and, on the other, to promoting the competitiveness of European market players facing the dynamism of the United States and Japan, who were more advanced in their process of market deregulation [2, 3].

The goal of the Community was twofold: to deregulate the telecommunications market, and to promote the harmonization of national regulations.

The Commission had already proposed: a phased opening of the terminal equipment market; the acceptance by national telecommunications administrators of the obligation to interconnect and provide access to transfrontier service providers; and a clear separation of regulatory and operational functions.

The next steps involved a common definition of an agreed set of conditions for open network provision (ONP) to service providers and users, promoting objectives such as European-wide compatibility and interoperability between services provided within the common market.

In the next paragraphs we will briefly analyze the regulations that were enacted to achieve these goals.

5.2.1 The liberalization period

In 1988, the Commission Directive 88/301/EEC of 16 May on competition in the markets in telecommunications terminal equipment was adopted. The State monopoly at that time extended not only to the use of telecommunications networks, but also to the supply of terminal equipment. Liberalizing the latter market was the first step in a smooth transition to full competition.

The Commission Directive 90/388/EEC of 28 June 1990 on competition in the markets for telecommunications services ('Services Directive') began, for the first time, to address the principal issue: 'Member States shall withdraw all special or exclusive rights for the supply of telecommunications services other than voice telephony and shall take the measures necessary to ensure that any operator is entitled to supply such telecommunications services' (article 2).[1]

If the State made the supply of telecommunications services subject to license or authorization, it would 'ensure that the conditions for the grant of licences are objective, nondiscriminatory and transparent, that reasons are given for any refusal, and that there is a procedure for appealing against any such refusal.'

The directive applied only to the so-called value-added services, defined as 'services whose provision consists wholly or partly in the transmission and routing of signals on the public

---

[1] Telex, mobile radiotelephony, paging and satellite services were not regulated by the Directive (article 1 (2)).

telecommunications network by means of telecommunications processes, with the exception of radio broadcasting and television' (article 1). Voice telephony was excluded, together with telex, mobile radiotelephony and satellite services.

Value-added telecommunications services were now included in the competition framework set up by article 59 of the Treaty (now article 49, on the freedom to provide services within the Community) in conjunction with article 86 (now article 82, on the prohibition of abuse of a dominant position within the common market or substantial part of it): granting exclusive or special rights for the supply of those services was deemed a violation of article 90 (now article 86).

The Council Resolution of 22 July 1993 'on the review of the situation in the telecommunications sector and the need for further development in that market'[2] in recognizing a general requirement 'for maintaining the financial stability of the sector and safeguarding universal service', considered the liberalization of all public voice telephony services, 'whilst maintaining universal service', a major goal for the Community's telecommunications policy 'in the longer term', to be achieved by 1998.

In 1994, the Satellite Directive[3] amended article 2 of Directive 90/388/EEC, by deepening the obligation of Member States in the liberalization process. They should now 'take the measures necessary to ensure that any operator is entitled to supply any such telecommunications services, other than voice telephony'. This directive also extended the 'Services Directive' to satellite services.

Directive 95/51/CE[4] extended the discipline of the 'Services Directive' to telecommunications services provided through cable television networks. This discipline was also extended to mobile and personal communication services in Directive 96/2/EC.[5]

The last step was to extend the competition to voice telephony. That happened with Commission Directive 96/19/EC of 13 March 1996, which amended Directive 90/388/EEC with regard to the implementation of full competition in telecommunications markets.

5.2.2 The harmonization effort

During the liberalization period, the Community enacted a set of directives to harmonize national regulations in their transition towards a deregulated market. Since harmonization and liberalization occurred at the same time, the NRF was initially directed at public or private bodies that were accorded exclusive or special rights in the provision of telecommunications services by Member States. The goal of such a transitory homogeneous legal framework was to provide a set of rules to enable new undertakings to enter a strongly monopolistic, or oligopolistic market. In order to achieve this goal, some principles for access to public telecommunications networks were necessary.

---

[2] O. J. 06/08/1993, C213/1.

[3] Commission Directive 94/46/EC of 13 October 1994 amending Directive 88/301/EEC and Directive 90/388/EEC, in particular with regard to satellite communications.

[4] Commission Directive 95/51/EC of 18 October 1995 amending Directive 90/388/EEC with regard to the abolition of the restrictions on the use of cable television networks for the provision of already liberalized telecommunications services.

[5] Commission Directive 96/2/EC of 16 January 1996 amending Directive 90/388/EEC with regard to mobile and personal communications. Mobile and personal communication services are defined as 'services whose provision consists, wholly or partly, in the establishment of radiocommunications to a mobile user, and makes use wholly or partly of mobile and personal communications systems' (article 1).

The ONP Framework Directive,[6] adopted in conjunction with the 'Services Directive', was aimed at 'the harmonization of conditions for open and efficient access to, and use of, public telecommunications networks and, where applicable, public telecommunications services' (article 1).

The ONP conditions, defined as the conditions 'which concern the open and efficient access to public telecommunications networks and, where applicable, public telecommunications services and the efficient use of those networks and services', should comply with some basic principles[7] and can be restricted only for reasons based on essential requirements,[8] such as security of network operations, maintenance of network integrity, interoperability of services, and protection of data (article 3 (2)).

This directive established the procedure and timetable of the harmonization effort and represented the general framework of the regulations to come, which were designed to create a uniform legal environment in all Member States for telecommunication services.

The first directive enacted within this framework was the ONP Leased Lines Directive,[9] whose goal was 'the harmonization of conditions for open and efficient access to, and use of, the leased lines provided to users on public telecommunications networks, and the availability throughout the Community of a minimum set of leased lines with harmonized technical characteristics' (article 1).

Leased lines are, according to the definition provided by article 2, 'the telecommunications facilities provided in the context of the establishment, development and operation of the public telecommunications network, which provide for transparent transmission capacity between network termination points, and which do not include on-demand switching'.

Article 3 imposed a general duty on Member States to provide information on technical characteristics, tariffs, supply and usage conditions of leased lines. According to article 6, restrictions on the use of leased lines should be aimed only 'at ensuring compliance with the essential requirements', defined by the ONP Framework Directive, and imposed 'by the national regulatory authorities through regulatory means'.

Moreover, Member States should guarantee a minimum set of leased lines, with defined technical requirements, throughout the Community (article 7).

In 1995, Directive 95/62/EC extended the ONP framework to voice telephony.[10] The directive established principles to regulate access, interconnection, tariffs, transparency, the obligation to inform, and the access and use of fixed public telephone networks.

The problem concerning the procedures for granting authorization to provide telecommunications services was addressed by Directive 97/13/EC.[11] According to article 3(2), and as

---

[6] Council Directive 90/387/EEC of 28 June 1990 on the establishment of the internal market for telecommunications services through the implementation of open network provision.

[7] Article 3 states that: they must be based on objective criteria; they must be transparent and published in an appropriate manner; and they must guarantee equality of access and must be nondiscriminatory, in accordance with Community law.

[8] Essential requirements are, according to article 2 (6), 'the non-economic reasons in the general interest which may cause a Member State to restrict access to the public telecommunications network or public telecommunications services. These reasons are security of network operations, maintenance of network integrity and, in justified cases, interoperability of services and data protection. Data protection may include protection of personal data, the confidentiality of information transmitted or stored, as well as the protection of privacy.'

[9] Council Directive 92/44/EEC of 5 June 1992 on the application of open network provision to leased lines.

[10] Directive 95/62/EC of the European Parliament and of the Council of 13 December 1995 on the application of open network provision (ONP) to voice telephony.

[11] Directive 97/13/EC of the European Parliament and of the Council of 10 April 1997 on a common framework for general authorizations and individual licences in the field of telecommunications services.

a general principle, 'Member States shall ensure that telecommunications services and/or telecommunications networks can be provided either without authorization, or on the basis of general authorizations'. If an undertaking complies with the conditions attached to a general authorization- the conditions are to be 'published in an appropriate manner so as to provide easy access to that information for interested parties' (article 4)-, Member States cannot prevent it from providing the intended telecommunications services (article 5).

With Directive 97/33/EC,[12] the problem of interconnection, universal service and interoperability of networks was addressed.

In 1998, Directive 95/62/EC was replaced by Directive 98/10/EC and added a regulatory framework to the liberalized voice telephony market.[13] Indeed, according to Recital (3), 'in moving to a competitive market, there are certain obligations which should apply to all organizations providing telephone services over fixed networks, whereas there are others which should apply only to organizations enjoying significant market power or which have been designated as a universal service operator'. That required substantial modifications to Directive 95/62/EC.

5.3 The New Regulatory Framework

In 2002, the European Parliament and the Council adopted the New Regulatory Framework (NRF), replacing the Services and ONP Directives, because the preceding 'regulatory framework for telecommunications has been successful in creating the conditions for effective competition in the telecommunications sector during the transition from monopoly to full competition.' The objective of the new framework was to simplify and rationalize the existing regulations, developed over a decade of rapid change. New technologies had evolved and the market was now populated by very different players [4].

The NRF consists of a set of directives[14] to replace the ONP Directives, a new Liberalization Directive[15] which replaces the Services Directive, as amended by successive directives, and a directive on privacy and electronic communications.[16]

The scope of the new regulations was significantly wider: whereas the previous regulatory framework concerned telecommunications services and networks, the new

---

[12] Directive 97/33/EC of the European Parliament and of the Council of 30 June 1997 on interconnection in Telecommunications with regard to ensuring universal service and interoperability through application of the principles of open network provision (ONP).

[13] Directive 98/10/EC of the European Parliament and of the Council of 26 February 1998 on the application of open network provision (ONP) to voice telephony and on universal service for telecommunications in a competitive environment.

[14] Directive 2002/21/EC of the European Parliament and of the Council of 7 March 2002 on a common regulatory framework for electronic communications networks and services (Framework Directive).

Directive 2002/19/EC of the European Parliament and of the Council of 7 March 2002 on access to, and interconnection of, electronic communications networks and associated facilities (Access Directive).

Directive 2002/20/EC of the European Parliament and of the Council of 7 March 2002 on the authorization of electronic communications networks and services (Authorization Directive).

Directive 2002/22/EC of the European Parliament and of the Council of 7 March 2002 on universal service and users' rights relating to electronic communications networks and services (Universal Service Directive).

[15] Commission Directive 2002/77/EC of 16 September 2002 on competition in the markets for electronic communications networks and services.

[16] Directive 2002/58/EC of the European Parliament and of the Council of 12 July 2002 concerning the processing of personal data and the protection of privacy in the electronic communications sector (Directive on privacy and electronic communications)

framework encompassed the broader notion of 'electronic communications' services and networks.[17]

Article 2(c) of the Framework Directive defines the meaning of 'electronic communications service' as a 'service normally provided for remuneration which consists wholly or mainly in the conveyance of signals on electronic communications networks, including telecommunications services and transmission services in networks used for broadcasting'. This definition specifically excludes 'services providing, or exercising editorial control over, content transmitted using electronic communications networks and services'.

The NRF came into force on 25 July 2003.

5.3.1 Liberalization

The new Liberalization Directive takes into account the latest technological developments and 'the convergence phenomenon which has shaped the information technology, media and telecommunications industries over recent years.'[18] Its definition of electronic communications services and networks is broader than previous definitions, and includes the transmission and broadcasting of radio and television programs,[19] fiber, fixed and mobile telecommunications networks, as well as cable television networks and those used for terrestrial broadcasting, along with satellite and Internet, whether used for voice, fax, data or images.

Article 2 of the directive prohibits Member States from 'grant(ing) or maintain(ing) in force exclusive or special rights for the establishment and/or the provision of electronic communications networks, or for the provision of publicly available electronic communications services.' Moreover, they 'shall take all measures necessary to ensure that any undertaking is entitled to provide electronic communications services or to establish, extend or provide electronic communications networks.'

The conditions attached to general authorization 'shall be based on objective, nondiscriminatory, proportionate and transparent criteria.'

Article 4 prohibits the awarding of special or exclusive rights for the use of radio frequency, and requires that their assignment be based 'on objective, nondiscriminatory, proportionate and transparent criteria'. The same criteria must be applied to the design of the national scheme, so that the cost of the provision of universal service obligations is shared among telecommunications organizations.

---

[17] According to article 2(a) of the Framework Directive, electronic communication networks are 'transmission systems and, where applicable, switching or routing equipment and other resources which permit the conveyance of signals by wire, by radio, by optical or by other electromagnetic means, including satellite networks, fixed (circuit and packet switched, including Internet) and mobile terrestrial networks, electricity cable systems, to the extent that they are used for the purpose of transmitting signals, networks used for radio and television broadcasting, and cable television networks, irrespective of the type of information conveyed.'

[18] Directive 2002/77/EC, Recital (6).

[19] According to Recital (8), 'special or exclusive rights which amount to restricting the use of electronic communications networks for the transmission and distribution of television signals are contrary to Article 86(1), read in conjunction with Article 43 (right of establishment) and/or Article 82(b) of the EC Treaty insofar as they have the effect of permitting a dominant undertaking to limit 'production, markets or technical development to the prejudice of consumers.'

5.3.2 National Regulatory Authorities

The Framework Directive defines the role and tasks of the National Regulatory Authorities (NRAs) and their regulatory powers.

NRA independence must be guaranteed 'by ensuring that they are legally distinct from, and functionally independent of, all organizations providing electronic communications networks, equipment or services' (article 3 (2)). NRAs and national competition authorities must cooperate through the exchange of information whenever this is required for the application of the provisions of the New Regulatory Framework (article 3 (5)).

One of the major tasks of the NRAs is to 'contribute to the development of the internal market by cooperating with each other and with the Commission in a transparent manner to ensure the consistent application, in all Member States, of the provisions of this Directive and the Specific Directives' (article 7 (2)). They must promote competition, ensuring access and an efficient use of resources in the provision of electronic communications services and networks, and the interests of 'the citizens of the European Union', ensuring access to a universal service and a high level of consumer and privacy protection (article 8).

The NRAs have a role in the management of radio frequency for electronic communications services. They are responsible for spectrum assignment.

Article 12 of the Framework Directive addresses the problem of sharing facilities. On the one hand, each undertaking providing communications services 'has the right under national legislation to install facilities on, over or under public or private property, or may take advantage of a procedure for the expropriation or use of property', on the other, the NRAs must 'encourage the sharing of such facilities or property' (article 12 (1)). In addition, 'if undertakings are deprived of access to viable alternatives because of the need to protect the environment, public health, public security or to meet town and country planning objectives, Member States may impose the sharing of facilities or property (including physical colocation) on an undertaking operating an electronic communications network', provided that the arising costs are shared too.

According to article 20, the NRAs have a role in dispute resolution between undertakings, and the authority to issue binding decisions 'in the shortest possible timeframe and in any case within four months except in exceptional circumstances'.

Any user or undertaking 'who is affected by a decision of a national regulatory authority has the right of appeal against the decision to an appeal body that is independent of the parties involved. This body, which may be a court, shall have the appropriate expertise available to it to enable it to carry out its functions' (article 4).

An important part of the Framework Directive is dedicated to the promotion of effective competition. In order to achieve this goal, special provisions regarding undertakings with significant market power are laid down in articles 14–16 (see [5–9]).

According to article 14 (2), 'an undertaking shall be deemed to have significant market power if, either individually or jointly with others, it enjoys a position equivalent to dominance, that is to say, a position of economic strength affording it the power to behave to an appreciable extent independently of competitors, customers and ultimately consumers.' If a Member State's NRA determines that a relevant market is not effectively competitive 'it shall identify undertakings with significant market power on that market in accordance with Article 14 and [...] shall on such undertakings impose appropriate specific regulatory obligations' (*ex ante* regulation, see article 16 (4)) for access and interconnection (articles 7 and 8 of the Access Directive), or to inhibit anticompetitive practices (articles 16, 17, 18 and 19 of the Universal Service Directive).

In 2002, the Commission adopted guidelines to standardize the procedure of market analysis for determining significant marker power[20] and a recommendation for the definition of product and service markets, 'the characteristics of which may be such as to justify the imposition of regulatory obligations' set out by the other directives of the NRF (article 15(1)).[21]

5.3.3 General authorization

The aim of the Authorization Directive is 'to implement an internal market in electronic communications networks and services through the harmonization and simplification of authorization rules and conditions in order to facilitate their provision throughout the Community' (article 1).

As stated in Recital (3), the main goal of the directive is to 'ensure the freedom to provide electronic communications networks and services'. To achieve this goal, the 'least onerous authorization system possible should be used to allow the provision of electronic communications networks and services, in order to stimulate the development of new electronic communications services and pan-European communications networks and services, and to allow service providers and consumers to benefit from the economies of scale of the single market'.[22] As a result, a general authorization for the provision of electronic communications services and networks was applied 'without requiring any explicit decision or administrative act by the national regulatory authority, and by limiting any procedural requirements to notification only'.[23]

According to article 3(2), '[t]he provision of electronic communications networks or the provision of electronic communications services may [. . .] only be subject to a general authorization. The undertaking concerned may be required to submit a notification but may not be required to obtain an explicit decision or any other administrative act by the national regulatory authority before exercising the rights stemming from the authorization. Upon notification, when required, an undertaking may begin activity'.

Rights arising from the general authorization include the right to provide electronic communication networks and services, and the right to install the facilities required in accordance with article 11 of the Framework Directive (article 4(1)). If an undertaking provides these services and networks to the public, it also has the right to negotiate interconnection with, and access to, publicly available networks and services of other providers, and the opportunity to become the designated provider of elements of a universal service (article 4(2)).

Conditions, 'objectively justified in relation to the network or service concerned, nondiscriminatory, proportionate and transparent', may be attached to the general authorization (article 6). Moreover, the authorization may impose specific obligations relating to access and universal service.

---

[20] Commission guidelines on market analysis and the assessment of significant market power under the Community regulatory framework for electronic communications networks and services, O.J. 11/07/2002, C165/06. The guidelines are required by article 15 (2) of the Framework Directive.

[21] Commission Recommendation of 11 February 2003 on relevant product and service markets within the electronic communications sector susceptible to *ex ante* regulation in accordance with Directive 2002/21/EC of the European Parliament and of the Council on a common regulatory framework for electronic communication networks and services, O.J. 08/05/2003, L114/45.

[22] Directive 2002/20/EC, Recital (7)

[23] Directive 2002/20/EC, Recital (8).

Slightly different rules govern the allocation of radio frequency. Article 5(1) states that 'Member States shall, where possible, in particular where the risk of harmful interference is negligible, not make the use of radio frequencies subject to the grant of individual rights of use, but shall include the conditions for usage of such radio frequencies in the general authorization.' That means that the general rule may require, for the use of radio spectrum, individual rights to be granted by the State.

The objective of Decision 676/2002/EC[24], is to establish 'a policy and legal framework in the Community in order to ensure the coordination of policy approaches and, where appropriate, harmonized conditions with regard to the availability and efficient use of the radio spectrum' (article 1), and only requires Member States to publish information on rights, conditions, procedures, charges and fees concerning the use of radio spectrum (article 5).

5.3.4 Access and interconnection

The Access Directive applies to the provision of publicly available electronic communications services. Non-public networks and services providing content are not covered by it.

The aim of the directive is to 'establish a regulatory framework, in accordance with internal market principles, for the relationships between suppliers of networks and services that will result in sustainable competition, interoperability of electronic communications services and consumer benefits' (article 1) [10, 11]. Indeed, as Recital (6) points out, '[i]n markets where there continue to be large differences in negotiating power between undertakings, and where some undertakings rely on infrastructure provided by others for delivery of their services, it is appropriate to establish a framework to ensure that the market functions effectively.' This is why 'National Regulatory Authorities should have the power to secure, where commercial negotiation fails, adequate access and interconnection and interoperability of services in the interest of end users. In particular, they may ensure end-to-end connectivity by imposing proportionate obligations on undertakings that control access to end users.'

According to article 4, '[o]perators of public communications networks shall have a right and, when requested by other undertakings so authorized, an obligation to negotiate interconnection with each other for the purpose of providing publicly available electronic communications services, in order to ensure provision and interoperability of services throughout the Community.' Article 5 regulates the powers of NRAs to encourage and, where appropriate ensure, access, interconnection and interoperability of services 'in a way that promotes efficiency, sustainable competition, and gives the maximum benefit to end users.' More specifically, the NRAs must have the power to impose obligations of transparency, 'requiring operators to make public specified information, such as accounting information, technical specifications, network characteristics, terms and conditions for supply and use, and prices' (article 9), and nondiscrimination, thereby ensuring that 'the operator applies equivalent conditions in equivalent circumstances to other undertakings providing equivalent services, and provides services and information to others under the same conditions and of the same quality as it provides for its own services, or those of its subsidiaries or partners' (article 10).

NRAs have the power to impose on operators obligations to share facilities (article 12) and obligations 'relating to cost recovery and price controls, including obligations for cost

---

[24] Decision No 676/2002/EC of the European Parliament and of the Council of 7 March 2002 on a regulatory framework for radio spectrum policy in the European Community (Radio Spectrum Decision), O.J. 24/04/2002, L108 /1.

orientation of prices and obligations concerning cost accounting systems, for the provision of specific types of interconnection and/or access, in situations where a market analysis indicates that a lack of effective competition means that the operator concerned might sustain prices at an excessively high level, or apply a price squeeze, to the detriment of end users' (article 13).

5.3.5 Universal service

The problem of universal service was one of the major obstacles in the liberalization process, especially in voice telephony. The Universal Service Directive (USD) addresses this issue.

The USD and compensation can have a deep impact on a competitive market. Indeed, as stated by Recital (4) of the directive, 'the provision of a defined minimum set of services to all end users at an affordable price may involve the provision of some services to some end users at prices that depart from those resulting from normal market conditions. However, compensating undertakings designated to provide such services in such circumstances need not result in any distortion of competition, provided that designated undertakings are compensated for the specific net cost involved and provided that the net cost burden is recovered in a competitively neutral way.' [12, 13]

The aim of the directive is to 'ensure the availability throughout the Community of good quality publicly available services through effective competition and choice and to deal with circumstances in which the needs of end users are not satisfactorily met by the market' (article 1(1)). In order to achieve these goals, the directive, 'establishes the rights of end users and the corresponding obligations on undertakings providing publicly available electronic communications networks and services', thus defining a 'minimum set of services of specified quality to which all end users have access, at an affordable price in the light of specific national conditions, without distorting competition' (article 1(2)).

According to article 4, 'Member States shall ensure that all reasonable requests for connection at a fixed location to the public telephone network and for access to publicly available telephone services at a fixed location are met by at least one undertaking'. The connection should allow end users 'to make and receive local, national and international telephone calls, facsimile communications and data communications, at data rates that are sufficient to permit functional Internet access' (article 4(2)). The directive does not require a specific amount of bandwidth to be provided. Recital (7) clearly states that universal service is not extended to Integrated Service Digital Network (ISDN), and requires Member States to take 56 kbits of data rate, available with voice band modems, as a result to be achieved.

One or more undertakings can be designated to guarantee the provision of universal service. Designation must be carried out with an 'efficient, objective, transparent and nondiscriminatory designation mechanism, whereby no undertaking is *a priori* excluded from being designated' (article 8).

The costs of universal service are calculated by the NRAs with parameters specified in article 12. Where they find that 'an undertaking is subject to an unfair burden' because of universal service provision, Member States, if requested by the designated undertaking, can transparently compensate it with public funds and/or share the net cost of universal service between electronic communications services and network providers (article 13).

End users' rights and interests are regulated by articles 20–31, and include transparency on prices and tariffs, quality of service requirements, interoperability of digital television equipment, operator assistance and directory services, the single European call number (112), and number portability between operators, etc.

Article 18 deals with the competitiveness of the leased lines market. If an NRA determines that competition is not effective, it can identify undertakings with significant market power and require them to provide a minimum set of leased lines.

5.3.6 Radio local area networks

We have seen that, according to article 5(1) of the Authorization Directive, where possible the use of radio frequencies should not be subject to the granting of individual rights of use, particularly 'where the risk of harmful interference is negligible', but to general authorization. In 2003, the European Commission adopted a Recommendation to encourage this approach for radio local area networks (R-LANs) too.[25]

Wireless Internet connectivity is an innovative technology with the potential for serious impact in the provision of community network services, especially in less advantaged locations. It uses frequencies included in the 2400.0–2483.5 MHz (2.4 GHz band) or the 5150–5350 MHz or 5470–5725 MHz bands (5 GHz bands).

The Recommendation states that 'Member States should allow the provision of public R-LAN access to public electronic communications networks and services in the available 2.4 GHz and 5 GHz bands to the extent possible without sector specific conditions and in any case subject only to general authorization.'

5.3.7 E-privacy

The provision of electronic communications services and networks has a direct impact on a key legal issue of the digital age: the right to privacy [14].

In 1995, an important European directive laid down the principles for the protection of individuals with regard to the processing of personal data.[26] Those principles were translated into specific rules for the telecommunication sector in 1997.[27]

The advances in digital technologies, especially the deployment of digital networks and their large capacities and possibilities for processing personal data, necessitated the replacement of the 1997 directive, with Directive 2002/58/EC of the European Parliament and the Council of 12 July 2002. It covers personal data processing and the protection of privacy in the electronic communications sector (Directive on privacy and electronic communications).

Its aim is to 'harmonize the provisions of the Member States required to ensure an equivalent level of protection of fundamental rights and freedoms, and in particular the right to privacy, with respect to the processing of personal data in the electronic communication sector and to ensure the free movement of such data and of electronic communication equipment and services in the Community' (article 1).

---

[25] Commission Recommendation of 20 March 2003 on the harmonization of the provision of public R-LAN access to public electronic communications networks and services in the Community, O. J. 25/03/2003, L078/12.

[26] Directive 95/46/EC of the European Parliament and of the Council of 24 October 1995 on the protection of individuals with regard to the processing of personal data and on the free movement of such data.

[27] Directive 97/66/EC of the European Parliament and of the Council of 15 December 1997 concerning the processing of personal data and the protection of privacy in the telecommunications sector.

According to article 3, the E-Privacy Directive applies 'to the processing of personal data in connection with the provision of publicly available electronic communications services in public communications networks in the Community.'

The directives establish some duties for the electronic communications services and network provider in relation to:

1. The security of their infrastructures and the measures to be taken, 'having regard to the state of the art and the cost of their implementation', to 'ensure a level of security appropriate to the risk presented' (article 4).
2. Processing traffic data. This must be 'erased or made anonymous when it is no longer needed for the purpose of the transmission of a communication', and processing can occur only to the extent 'necessary for the purposes of subscriber billing and interconnection payments' (article 6).
3. Processing location data other than traffic data is possible only when data 'are made anonymous, or with the consent of the users or subscribers to the extent and for the duration necessary for the provision of a value-added service' (article 9).

Article 13 deals with unsolicited communications, stating that 'the use of automated calling systems without human intervention (automatic calling machines), facsimile machines (fax) or electronic mail for the purposes of direct marketing may only be allowed in respect of subscribers who have given their prior consent.'

Member States are required to enact legislation to ensure the confidentiality of communications. In particular, 'they shall prohibit listening, tapping, storage or other kinds of interception or surveillance of communications and the related traffic data by persons other than users, without the consent of the users concerned, except when legally authorized' (article 5).

Legislative measures to restrict the scope of the rights and obligations laid down by the directive may be adopted if the restriction 'constitutes a necessary, appropriate and proportionate measure within a democratic society to safeguard national security (i.e. State security), defence, public security, and the prevention, investigation, detection and prosecution of criminal offences or of unauthorized use of the electronic communication system' (article 15(1)).

5.4 Member States' implementation: the Italian case

5.4.1 The Italian Electronic Communications Code

Italy implemented the NRF with the adoption of an Electronic Communication Code (ECC), enacted with Decreto Legislativo 1 Agosto 2003, n. 259.

The ECC is an attempt to provide a comprehensive and coherent regulation of the entire electronic communications sector. The basic principles, set out in article 4, can be summarized as:

- the freedom of communications;
- the secrecy and confidentiality of communications to be achieved through integrity and security of electronic communications networks; and
- the freedom of economic undertaking in the provision of electronic communications services and networks.

The Code applies to the provision of electronic communications services and networks for public and private use, except in the case of services providing, or implying editorial control over, content (article 2).

The main objectives include:

- the simplification of administrative procedures for the provision of electronic communications services and networks;
- universal service provision, with limited impact over competition;
- ensured competition in the provision of communication services, including broadband connectivity;
- the guaranteed convergence and interoperability of networks and services through the adoption of open standards; and
- guaranteed technological neutrality, without discriminating against any particular technology (article 4).

According to article 6, the State, Regions and other public local authorities are not allowed to directly provide publicly available electronic communications networks and services, unless through controlled undertakings or companies.[28]

5.4.1.1 Autorità per le Garanzie

The Italian NRA is the Autorità per le Garanzie nelle comunicazioni (article 1(1) f. and article 7(2) ECC).[29] The Autorità per le Garanzie, the Ministry of Communicaions and the Competition Authority (Autorità garante della concorrenza ed il mercato) must cooperate to ensure the application of European directives on electronic communications (article 8, Electronic Communications Code).

Article 9 regulates the right of appeal against the decisions of the Autorità per le Garanzie and the Ministry of Communications.

International cooperation, objectives and principles of the regulatory activities are set out in articles 12 and 13.

The Autorità per le Garanzie determines the effectiveness of competition in electronic communications markets, identifies undertakings with significant market power and imposes on them appropriate specific regulatory obligations (articles 17–19). It also has a role in dispute resolution between undertakings (article 23).

5.4.1.2 General authorization

The general authorization is regulated by article 25 of the ECC. The provision of electronic communications services and networks is subject to notification – to the Ministry of Communication – of the intention to start providing services. After notification, the operator can

---

[28] Among the Member States that implemented the New Regulatory Framework, the United Kingdom adopted a different approach. According to Section 349(2) of the Communications Act 2003, for the purpose of broadcasting or distributing information relating to their activities, local authorities may 'provide an electronic communications network or electronic communications service, or arrange with the provider of such a network or service for the broadcasting or distribution of such information by means of the network or service.'

[29] The Autorità di Garanzie nelle comunicazioni was created in 1997 with Legge 31 Luglio 1997, n. 249.

begin its activities. Within sixty days, the Ministry must verify the presence of the essential requirements for the provision of services, and fulfillment of the conditions attached to the general authorization, as laid down in article 28. If the operator cannot satisfy these conditions, the Ministry can order it to cease supplying services with a motivated order.

The use of radio frequencies 'where the risk of harmful interference is negligible', in accordance with article 5 of the Authorization Directive, is not subject to the granting of individual rights of use, only to general authorization (article 27(1)).

If the grant of rights is deemed necessary, the number of rights of use of radio frequencies can be limited, in order to ensure an efficient use of radio spectrum (article 27(6)), with the procedure laid down in article 29.

5.4.1.3 Access and interconnection

Access and interconnection are regulated by articles 40–52 of the Code. The principles established in these articles are applied directly from the rules of the Access Directive, and the wording of the provisions is almost identical to that of the European legislation. Operators have a right, and if requested an obligation, to negotiate access and interconnection (article 41). The Autorità per le Garanzie must ensure access, interconnection and interoperability of services in order to promote economic efficiency, sustainable competition and give the maximum benefit to end users (article 42).[30]

5.4.1.4 Universal service

Universal service is regulated in accordance with the Universal Service Directive. It includes a number of services that must be provided to all end users in the national territory: connection at a fixed location to the public telephone network (article 54), directory services (article 55), public pay telephones (article 56), and special measures for disabled users (article 57).

The Autorità per le Garanzie has the power to:

- designate undertakings for the provision of universal service in accordance with article 58;
- monitor the affordability of tariffs (article 59) and quality of services (article 61)
- calculate the costs of universal service provision (article 62)
- compensate the costs of universal service (article 63).

5.4.2 Internet service provision

Internet service provision was regulated for the first time by Decreto Legislativo 17 Marzo 1995, n. 103. It implemented the Services Directive (90/388/EC) and required preliminary authorization of the Ministry of Communications for Internet service providers to commence activity (article 3(3)). That rule changed in 1997 with the adoption of D.P.R. 19 Settembre 1997, n. 318, which extended the general authorization regime to all telecommunications services and networks, with the exception of voice telephony (article 6).

With the adoption of the ECC, Internet service provision is included in the general framework of electronic communications service and network provision.

---

[30] See article 5 of the Access Directive.

5.4.3 Radio local area networks

R-LAN deployment for private use was liberalized in 2001 with D.P.R. 5 Ottobre 2001, n. 447. According to article 6, their private use is totally free and not subject to the general authorization.

The provision of public R-LAN access is regulated by Decreto Ministero delle Comunicazioni 28 Maggio 2003. However, the use of frequencies in the 2400.0–2483.5 MHz (2.4 Ghz), 5001–5350 Mhz, and 5470–5725 Mhz (5 Ghz) bands is subject to general authorization (article 3).

Article 6 determines the conditions attached to general authorization with regard to security, integrity and interoperability of networks, technical requirements of the infrastructure, use of access code for users' identification, and privacy and confidentiality of communications.

5.4.4 A case study: Provincia Autonoma di Trento and Informatica Trentina

In 1983, The Provincia Autonoma di Trento[31] created Informatica Trentina S.p.a., in order to develop an electronic information system to collect and process socio-economic data and to promote administrative automation in local authorities and municipalities within the province.[32]

Informatica Trentina S.p.a is a public company whose shareholders are, among others, the Provincia Autonoma di Trento, the Municipality of Trento, the Municipality of Rovereto and other local authorities.

In 1989, Informatica Trentina created TELPAT, a private electronic communication network to provide telecommunications services to local authorities and municipalities, public health facilities, public tourist offices and other such bodies. With TELPAT, Informatica Trentina deployed the Catalogo Bibliografico Trentino, an electronic bibliographic catalog with more than 150 public and private libraries connected to it and located in the province.

As a result of the liberalization of the electronic communications sectors, Informatica Trentina became an Internet service provider in 1999, when it created a portal and started offering Internet connectivity to students and teachers from schools within the province,[33] in order to promote computer literacy among students.

Informatica Trentina can be thought of as an example of community network provision that the deregulation and liberalization of a formerly monopolist and closed market eventually made possible.

5.5 Summary

This chapter provided a brief description of the European Telecommunication Law, as reshaped by the adoption of the New Regulatory Framework in 2002.

Starting in 1986, the national regulations in the telecommunications sector have been liberalized and harmonized by European Community Law, with the aim of building a common

---

[31] The Autonomous Province of Trento (Italy) is a public local authority with a constitutionally recognized status, similar to that of the Regioni (Regions). It has limited legislative powers. For some introductory notes on Italian Constitutional Law and the role of Regions, see [15] particularly p. 55.

[32] Legge Provinciale 6 Maggio 1980, n. 10, Istituzione di un sistema informativo elettronico provinciale.

[33] See http://www.vivoscuola.it

competitive market. The liberalization and harmonization process was characterized by the adoption of an interim regulatory framework, to provide basic common principles in the provision of communications services and to regulate access and interconnection to telecommunications infrastructures controlled by governmental bodies through monopolies and exclusive or special rights.

The liberalized market required a new coherent set of rules that were enacted with the NRF. We outlined the basic principles of the NRF with reference to the provision of community network services. The NRF requires European Member States to implement it through the adoption of specific regulations. The new Italian Telecommunications Code was analyzed in order to provide an example of such implementation.

References

1. European Commission (1987), *Towards a Dynamic European Economy, Green Paper on the development of the common market for telecommunication services and equipment*, COM (87) 290, June 1987.
2. Lazer, D. and Mayer Schonberger, V. (2002), Governing Networks: Telecommunication Deregulation in Europe and the United States, *Brooklyn Journal of International Law*, **27**, 819.
3. Burkert, H. (2002), The Post-Deregulatory Landscape in International Telecommunication Law: A Unique European Union Approach?, *Brooklyn Journal of International Law*, **27**, 739.
4. Garzaniti, L. (2003), *Telecommunications, Broadcasting and the Internet: EU Competition Law and Regulation*, Sweet & Maxwell, London.
5. de Streel, A. (2003), Market Definitions in the New European Regulatory Framework for Electronic Communications, *Info*, **5**, 27.
6. de Streel, A. (2003), The New Concept of 'Significant Market Power' in Electronic Communications: the Hybridization of the Sectoral Regulation by Competition Law, *European Competition Law Review*, **24**, 535.
7. Maxwell, W. (ed.) (2002), *Significant Market Power and Dominance in the Regulation of Telecommunications Markets*, New York.
8. Kahai, S. K., Kaserman, D. L. and Mayo, J. W. (1996), Is the 'Dominant Firm' Dominant? An Empirical Analysis of AT&T's Market Power, *Journal of Law and Economics*, **39**, 499.
9. Tarrant, A. (2000), Significant market power and dominance in the regulation of telecommunications markets, *European Competition Law Review*, **21**, 320.
10. Nikolinakos, N. T. (2001), The new European regulatory regime for electronic communications networks and associated services: the proposed framework and access/interconnection directives, *European Competition Law Review*, **22**, 93.
11. Ng, L.W. H. (1997), Access and interconnection issues in the move towards the full liberalization of European telecommunications, *The North Carolina Journal of International Law and Commercial Regulation*, **23**, 1.
12. Noam, E. M. (1997), Will Universal Service and Common Carriage Survive the Telecommunications Act of 1996?, *Columbia Law Review*, p. 955.
13. Frieden, R. M. (2000), Universal Service: When Technologies Converge and Regulatory Models Diverge, *Harvard Journal of Law and Technology*, **13**, 395.
14. Carey, P. and Ustaran, E. (2002), *E-privacy and online data protection*, Butterworths Law, London.
15. Comba, M. (2002), Constitutional Law, in J.S. Lena and U. Mattei, (eds) *Introduction to Italian Law*, The Hague, p. 31.

6

Models for Public Sector Involvement in Regional and Local Broadband Projects

Gareth Hughes
eris@, Belgium

6.1 Introduction

This chapter provides an overview of different models for deploying broadband and attempts to highlight the advantages and disadvantages of each approach. The choice of model depends on local circumstances, the legal situation of the country concerned, and also, to a large extent, on the will of the local or regional community. Therefore, after discussing the models, we are going to briefly summarize the different forms of public–private partnership. The third and last part of this chapter will provide some legal and economic considerations related to the models.

6.2 Overview of models for public involvement

Three of the five models are illustrated later by graphs in Figures 6.1 through 6.3, depicting the different levels in the broadband value chain. In each model, the level of intervention by the local or regional authority is marked with a horizontal dashed line. Tables 6.1 and 6.2, also later in the chapter, are intended to help in making comparisons between relative advantages and disadvantages, and in assessing how well the different models can be combined.

Regardless of the model that is chosen, there are some important steps to be taken:

1. Communities will need to raise awareness of the benefits of broadband services to all stakeholders of the community to stimulate interest.

Broadband Services: Business Models and Technologies for Community Networks.   Edited by I. Chlamtac,
A. Gumaste and C. Szabó   © 2005 John Wiley & Sons, Ltd. ISBN 0-470-02248-5.

2. They will need to carry out a 'community needs assessment' of both public and private sectors in order to estimate potential demand for broadband services.
3. They will need to establish the business case for whatever type of intervention they choose in order to ensure efficient usage of any public funds.
4. Furthermore, communities will have to decide whether they want to act on their own, or if they would prefer to enter into a public–private partnership. Section 6.4 of this chapter describes in more detail the most common forms of public–private partnership.
5. Finally, communities that decide to build a network or offer end user services will have to set up a separate legal entity at arm's length from the local authority. This is a legal requirement flowing from the EU regulatory framework for electronic communications, which requires that those authorities that have regulatory powers relevant to electronic communications services have to be separate from those entities that provide these services. The objective of this legal requirement is to ensure that there will be no conflict of interest for the public authority in its role as the authority awarding rights of way to various operators.

6.2.1 Community-operated networks and services

In this model, the local community builds, owns and operates a broadband network and provides services to end users. The resulting entity must acquire the appropriate 'authorization' (and eventually the right to use spectrum) from its national regulatory authorities, and must fully comply with the telecom sector's market rules and regulations. Note that a set of new EU telecom directives are being transformed into national laws in all EU Member States in 2004. An entity operating a broadband network will be defined as an 'undertaking offering electronic communication services over an electronic communication network' (see Chapter 5). At the same time, the obligation to apply for a telecommunication license before starting operations disappears – however, care has to be taken to comply with the obligations laid down in law for operators (so-called general authorization).

In this approach, regions or communities intervene in the lowest three levels of the value chain (Figure 6.1).

The advantage of this model is that it offers a complete solution to the broadband needs of a community by a single actor. This could be particularly attractive to isolated rural areas (e.g. a community of mountain villages) with a low level of telecommunications investment, or where residents and businesses are widely dispersed over a geographic area.

Disadvantages of this model are as follows.

1. This model risks a negative impact on competition in networks and services. The public sector risks delaying or discouraging the investment of any commercial operators in all three levels, namely passive infrastructure, networks and services.
2. The model imposes a financial risk on the public sector. It requires investment of public money.
3. It also requires the public sector to find some technical and commercial expertise and support. This disadvantage could be mitigated through the formation of a public–private partnership.

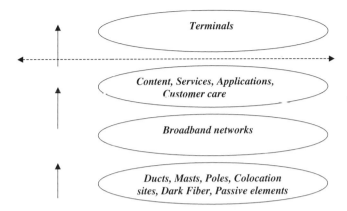

Figure 6.1　Community-operated networks and services

6.2.2 Carrier's carrier model

In this second model, communities build and operate broadband networks but lease 'raw band-width' to any commercial service provider interested in providing services via the community network. Here again, communities may do this on their own or through a public–private partner-ship. Service providers lease bandwidth to deliver their services to the end users and, possibly also, space in colocation centers for their own equipment.

In this model, the public sector intervenes in the two lowest levels of the value chain, see Figure 6.2. It could be preferable to create a separate entity to maintain and manage the passive infrastructure, independent of the network carrier.

The advantage of this model is that community investment in costly passive infrastruc-ture (trenches, ducts, etc.) and in broadband networks (switches, transmission equipment)

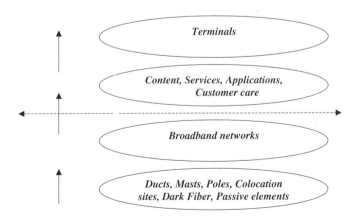

Figure 6.2　Carrier's carrier model

significantly lowers market entry costs for service providers and content providers, allowing them to diversify and extend their services to areas where it would otherwise have been prohibitively expensive to invest.

Disadvantages of the carrier's carrier model are as follows.

1. Because the public sector intervenes in the first two layers of the communications networks, there is thus a risk of discouraging any new private investment in these first two layers, or possibly of restricting competition in alternative networks and technologies.
2. There is a financial risk to the public sector.
3. The public sector needs to find technical expertise and support in order to build and manage the network (this disadvantage can be mitigated through public–private partnerships).

6.2.3 Passive infrastructure model

Passive infrastructure includes trenches, ducts, masts, manholes, colocation sites, dark fiber and other civil structures necessary for the deployment of broadband networks. In this third model, communities build the passive infrastructure that underlies broadband networks. They then lease some, or all, of the elements of the passive infrastructure to one or more operators, who complete the installation with their own network equipment. The maintenance and administration of the passive infrastructure is usually transferred to an independent third party, who takes care of leasing the passive infrastructure to operators.

In this model, the public sector intervenes only in the lowest level of the value chain. This is, nonetheless, the biggest entry barrier for any broadband operator. Passive infrastructure represents in the range of 70 % of the cost of a new fixed network, and often nearly 40 % of the cost of a wireless network.

Advantages of this model include most of the advantages of the carrier's carrier model. Furthermore, in this model the public sector intervenes only at the lowest level of the value chain, which can enhance competition in all higher levels of the value chain, such as networks, technologies, services and content (see Figure 6.3). This type of public intervention at such a low level is generally considered as competition neutral, and should not, in principle, cause any concerns regarding competition law.

The model shares the same disadvantages as the carrier's carrier model. Furthermore, it has one additional disadvantage. It can still require operators to make a sizeable investment in broadband networks (e.g. their own routing, switching or transmission equipment). The entry barriers for network operators and service providers, in particular, would be higher than in the carrier's carrier model, where the network is already activated and service providers need to make a very small initial investment (for example, into content and service development, marketing, etc.).

This need for additional investment creates a further risk for the local community managing the passive infrastructure. What if there is no operator or service provider willing to use the passive infrastructure that the community has built? Such a risk can, however, be mitigated through a reasonable business plan and a professional and realistic assessment of local needs.

Another possible way to address this problem is to create one independent entity to own and manage the passive infrastructure, and another to operate the broadband network. A clear structural separation between the passive infrastructure entity and the entity owning and managing

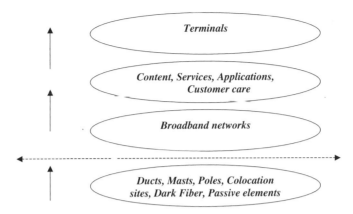

Figure 6.3   Passive infrastructure model

the broadband network not only safeguards competition neutrality, but it also ensures an anchor tenant (the local government) for the infrastructure from day one.

6.2.4 Fiber condominium

In most cases, passive infrastructure does not include the last segment of end user access lines from the curb to an apartment inside a multidwelling, or within a given housing or business area. In the fiber condominium model, the last part of the dark fiber infrastructure is owned by the end users themselves. The contractor advertises that they intend to build a condominium fiber network and offers early participants special pricing before construction begins. This allows the contractor to guarantee early financing for the project and to demonstrate to bankers and other investors that there are enough customers committed to the project.

The condominium fiber network is operated like a condominium apartment building. The individual owners of individual fiber strands can do whatever they want with them. They are free to carry any type of traffic and to terminate the fiber in any way they so choose. The company that installs the fiber network is responsible for overall maintenance and repairing the fiber in case of breaks, moves, additions or changes. The 'condominium manager' charges the owners of the individual strands of fiber a small maintenance fee, which covers maintenance and right of way costs.

It should be noted that similar types of condominium can be set up using any type of broadband access technology, for example fixed wireless access or Wi-Fi.

The advantages of this model are as follows.

1. This is an approach that follows demand. The end user infrastructure is built only for customers who have asked for it.
2. It can be an excellent complement to the carrier's carrier model or the passive infrastructure model. The infrastructure manager in both these models may offer end users the possibility to construct a final fiber drop to their premises if they are willing to pay the cost of this investment.

3. Fiber to the end user's premises can be a platform that allows a huge amount of advanced services, such as video-on-demand, quality videoconferencing, various video games, advanced medical and education services.

This model's disadvantage is that it only makes sense if end users buy into the condominium. Consequently, these services will only become available if there is a critical mass of end users requesting access to this type of infrastructure.

The different aspects of building and operating condominium fiber networks are dealt with in Chapter 7.

6.2.5 Aggregation of demand

The strategy in this case focuses on having a large enough critical mass of users to provide an incentive for building broadband network elements. The community aggregates together core user groups to provide a guaranteed level of demand, and thus a guaranteed minimum revenue stream to commercial broadband operators.

Municipalities can aggregate demand from:

- government and public entities. In most cases, this includes local government entities, schools and hospitals and other public bodies.
- the public sector, but also from the private sector, through agreements with businesses or even individual users through 'preregistration schemes'. Regional or local Internet service providers and major regional businesses are the obvious first targets for these agreements.

Frequently, aggregation of demand is used in parallel with another initiative, such as the creation of an open access carrier's carrier network or a primary infrastructure project. The means of doing this is by first carrying out a 'community needs assessment' of both public and private sectors. If public sector demand creates a sufficient critical mass, the contract is tendered out to interested broadband operators. If the public sector demand does not seem to create a critical mass, public and private sectors' needs may be pooled into the contract. If necessary, demand from neighboring towns or areas can be pooled into the same contract in order to achieve the necessary critical mass.

The advantage of this type of intervention is that it bears no financial risk for the public sector. It does not require any cash investment, and the public sector does not need to develop any particular technical expertise.

Furthermore, and in particular when used in combination with another method of intervention, it reduces the financial risk for the private sector, and allows a faster and more effective deployment of broadband services in the area.

Disadvantages of this model are as follows.

1. It requires considerable effort to coordinate everybody's needs and to pool them into one contract. There is a need for strong leadership during detailed negotiations, in order to reach a consensus among the various parties on the best technology, the best operator, the best structure of contract, etc.

2. Some public sectors have good reasons to aggregate their demand on a national or regional basis and enter into a contract with a single broadband provider. This can be relevant to the police, hospitals or some educational establishments that share libraries or databases. However, it undermines 'local demand aggregation' efforts, because a large part of the potential demand is removed from the pool.
3. Reliance on this method alone for stimulating broadband may have an adverse effect on competition and restrict customers' choice over carriers and service options over a considerable time.
4. This model improves the business case of one broadband access operator, but risks destroying the business case for all the others in the same area. If a considerable part of the community is 'locked in' to a contract with one operator, there is very little business opportunity for any competing operator. However, this problem can be addressed if aggregation of demand is being used in combination with one of the other models discussed earlier – such as the carrier's carrier model or the passive infrastructure model. A combination of these two models may address the weakness of both models in an optimal manner.

6.2.6 Summary of models

Each model presents advantages and disadvantages, these are summarized in Table 6.1.

Some of these models can be complementary to one another, while combinations of some of them may lead to problems in terms of unfair competition. Table 6.2 attempts to capture how different approaches fit together.

6.3 Financial profiles of different models

The broadband value chain includes a number of different products and services with very different returns on investment. Figure 6.4 shows the expected return of investment of the various levels of the value chain.

As the figure illustrates, primary infrastructure is, in reality, very much like real estate. It is a long-term investment that requires massive capital expenditure in the beginning, but has a very long lifecycle. The typical lifecycle of this primary infrastructure is at least 20 years. It can thus be structured as an investment with a 15–20 year pay-back period. This type of investment, however, requires low risk that can only be provided through a public–private partnership. It is important to note that this is the typical investment profile of a typical utility. 'Utility' generally means a company which provides a service in the 'public interest'. Utilities tend to be heavily regulated because they need to ensure a certain universality of their service, while maintaining relatively low profitability. In exchange, they enjoy public support, normally in the form of a concession or in the form of public investment if they are required to service highly unprofitable areas. Depending on the country and its traditions, utilities can be fully state-owned, fully privately owned or companies of mixed ownership. Utilities have a very particular 'financial profile'. They involve very big investments that can only be paid back in extended periods of time, typically 15 to 20 years. However, they are very low-risk financial propositions and therefore 'attractive' to particular investors interested in safe, long-term investments.

Table 6.1 Advantages and disadvantages of different models

Model/Intervention	Advantages	Disadvantages
Community-operated networks and services	Complete solution to broadband needs by a single actor	Negative impact on competition in networks and services
		Public sector needs to develop technical and commercial expertise
		Financial risk for public sector
Carrier's carrier model	Positive impact on competition in services and contents	Public sector needs to develop some technical and commercial expertise (but not necessary in case of public–private partnership)
	Allows risk to be shared with private sector	Potential negative impact on competition in technologies/networks
Passive infrastructure model	Positive impact on facilities-based competition between networks, services and contents	Private operators still need to make a sizeable investment in network equipment
	Separates the financial risks of a telecom network (hi-tech) from the infrastructure (real estate) and allows reduction of the cost of capital	Public sector needs to develop some technical and commercial expertise (but not necessary in case of public–private partnership)
	Allows risk to be shared with private sector	Potential negative impact on competition in technologies/networks
Fiber condominium	Investments are only made where there is demand	Requires a critical mass of end users with an interest in this type of infrastructure
Aggregation of demand	No financial risk for public sector	If used alone, negative impact on competition in network services and contents
	Little need for technical expertise	
	Allows better definition and prioritization of needs, and a balanced service to users	Complex to organize and manage

A broadband communications network is a 'high-tech' investment. It has a much shorter lifecycle, typically 5–7 years, but creates more 'added value' in people's lives and businesses – allowing much higher returns, but also involving much higher risk. This means that loans for this type of investment are typically much more expensive than loans for utilities.

An argument can thus be made to keep passive infrastructure separate from the active parts of the network in order to lower the overall cost of investment. Private investment can be attracted to create and operate the passive infrastructure through a public–private partnership. This could also reduce the entry barriers for network operators and ensure a more sustainable flow of private investment in the upper levels of the value chain.

This different financial profile also means that a passive infrastructure company would be able to buy existing infrastructure from existing operators, refinance it by applying a different risk, and then lease it back to them.

Table 6.2   Combination of different models   +: very good combination; −: cannot be combined; ?: possible but could raise competition issues

Community-operated networks and services	Carrier's carrier model	Passive infrastructure model	Demand aggregation	Fiber condominium	
	−	−	?	+	Community-operated networks and services
−		−	+	+	Carrier's carrier network
−	−		+	+	Passive infrastructure model
?	+	+		+	Demand aggregation
+	+	+	+		Fiber condominium

6.4 Public–private partnerships

In this chapter, we often refer to public–private partnerships. This is a term used to describe a number of very different legal and commercial relationships, often very different from one country to another. We try to explain here some of the most common forms of these partnerships. The choice of public–private partnership depends, among other things, on relevant laws and legal traditions of the different countries.

- A joint venture between public and private sector with joint control: the public sector and a private entrepreneur set up a jointly-owned legal entity. Both share the capital expenditure and the commercial risk, and they are equally represented in the board of the joint venture. However, the daily management is often entrusted to the private partner. The public sector often keeps the right to veto a limited number of strategic decisions.

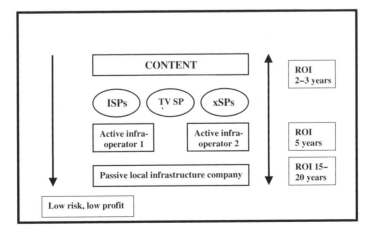

Figure 6.4   Expected ROI for different levels of the value chain

- A joint venture without joint control: the public sector and a private entrepreneur establish a company where the public sector maintains control, both at board level and management level.
- A direct subsidy or public sector loan or state guarantee: the public sector supports a private company in a particular investment by providing a subsidy or a state guarantee and lowering the risk, or by subsidizing the interest of the financing. The private sector company makes the capital expenditure and assumes all commercial risk.
- A management contract: the public sector builds a passive infrastructure and/or broadband network and enters into a management contract with a private undertaking. Management contracts often have durations of 3–5 years. In this case, it will be the public sector that invests the capital expenditure and assumes all commercial risk.
- A lease: the public authority builds a passive infrastructure and/or a broadband network and leases it to a private company for a duration of 10–15 years. In this case, the public sector makes the capital investment and retains ownership of the asset but shares the commercial risk with the private entrepreneur.
- A concession: the public sector awards a concession to a private company to carry out a project. This project may include the building and operation of a carrier's carrier network or passive infrastructure, or the operation of such a network or passive infrastructure. Concessions can take several forms. Some of the most common include:
  - BOT/BOO (build, operate, transfer/build, operate, own). The concessionaire finances the creation of the infrastructure/network and then exploits it for a number of years. After this initial period, ownership of the network/infrastructure can either revert to the public sector or remain with the concessionaire (depending on whether it is BOT or BOO).
  - The public sector finances and owns the assets but enters into a long-term contract with the private sector for operations, maintenance and investment.

6.5 Legal and economic considerations relating to models

In Chapter 5, the telecommunication regulatory environment in Europe was analyzed. In this section, we address some specific issues related to the public sector participation in providing telecommunication services.

Laws and market structures vary considerably from one country to another. As a result, it is not possible to provide a conclusive and comprehensive analysis of all legal and economic issues surrounding various types of intervention. Table 6.3 at the end of this chapter provides a summary of the legal situation. The above notwithstanding, it is useful to provide some fundamental considerations likely to apply to most, if not all, cases. A more detailed analysis of the specific circumstances remains necessary if a regional or local government decides to proceed to one, or a combination of, the methods of intervention discussed in this chapter. Many law firms offer specialized advice in the domains of European and national competition law.

Some of the fundamental state aid rules and competition rules in a nutshell are:

1. *State aid rules.* The Treaty on European Union prohibits any public subsidies to companies that distort competition in the European single market. Companies which received the subsidy would have a competitive advantage over their competitors. However, subsidies in less developed areas can be legal.

2. *Competition rules.* One of the most important objectives of the EU Treaty is to create a level playing field for all economic operators. In the telecom sector, European law prohibits the award of exclusive or special rights for telecommunication networks. As a general rule, local or national governments are prohibited from taking any action that would privilege one operator over another, as this could be deemed to constitute a special or exclusive right.

However, there are cases where local, regional or national authorities will have to make a choice and enter into some kind of relationship with some operator. This is legal, as long as the selection of the operator has been conducted through an open and transparent procedure, and in accordance with European and national rules on public procurement.

6.5.1 Aggregation of public sector demand

Aggregation of public sector demand may have a foreclosure effect on the market. There are a number of reasons.

First, public sector demand is likely to represent a sizeable share of the overall broadband market. If this demand is 'locked in' to one contract, the business case for any remaining operator will become difficult.

Legally speaking, a contract whereby parties agree that the broadband needs of the public sector will be handled by one operator over a certain period of time qualifies as an 'exclusive supply agreement'. In previous cases, the European Court of Justice took the view that exclusive supply agreements are likely to restrict competition if the following conditions are met:

- there are significant barriers to competitors entering the relevant market or increasing their market share;
- the agreement in question makes a significant contribution to those barriers.

(Case C-234/89 Delimitis [1991] ECR 935, [1992] CMLR 210, &27. Although stated in the context of a beer supply agreement, the principles set out in this judgment are of general application.)

The contract to serve aggregated public sector demand is awarded in accordance with public procurement laws. However, some of the public demand is likely to be better served through a ubiquitous and well-extended network, in which case, aggregation should not aggravate the disparities between operators.

6.5.2 Financial aid to an operator

Direct financial aid to an operator is likely to qualify as state aid. However, not all state aid is incompatible with the EU rules. For example, state aid can be compatible with EU rules if it qualifies under the guidelines on national or regional aid, or if the aid is considered necessary for promoting the economic development of areas 'where the standard of living is abnormally low, or there is serious under-employment' (a quote from the Treaty on European Union).

In other words, financial aid to operators in rural or under-served areas may be compatible with EU competition rules. However, it will be difficult to defend a case whereby financial aid is being granted to an operator in an urban or more developed area. The case for such aid becomes easier to make if the aid is subject to an obligation to operate the network on an open access basis and maintain a structural separation between the wholesale network operation activities and the retail service activities. Arguably, aid whereby the recipient receives financial assistance in order to create and maintain a network with open access to all operators and service providers enhances competition in services and therefore has no distortion effect. State aid with no distortion effect on competition is not illegal.

6.5.3 Community-operated networks and services

EU rules do not prevent municipalities or other public entities from becoming network operators. This is a matter of national law, which differs from one country to another. As a result, it goes far beyond the scope of this chapter to give a comprehensive analysis of the various national laws (but see the comparative Table 6.3).

However, there is one requirement under EU law that needs to be taken into consideration. Under the New Regulatory Framework that was adopted in December 2001, municipal network operators have to be set up at arm's length from the municipal authorities themselves, in order to ensure impartiality in awarding rights of way.

6.5.4 Carrier's carrier model

Similar considerations apply to municipal companies set up to create and operate a carrier's carrier network. EU rules do not prevent the creation of this type of company. National laws, however, may include some restrictions or limitations. Here again, these companies will need to be set up on an arm's length basis from the local authority itself, in order to avoid conflicts of interest in awarding rights of way.

Furthermore, this may be a model to consider in combination with financial aid to a private company, to create and operate such an open access network. As discussed above, financial aid to an open access network provider presents important advantages from a competition policy point of view. This type of financial aid would be clearly much easier to defend.

6.5.5 Passive infrastructure model

In a similar manner to the two cases described above, EU rules do not prevent the establishment of 'passive infrastructure companies' to create and operate passive infrastructure. However, national laws may provide some limitations. The obligation to set up this company at arm's length from the authority also applies in this case.

This model can also be considered in combination with financial aid. The deployment of passive infrastructure by a third party is likely to have a very positive impact on competition, since it promotes facilities-based competition.

Furthermore, this model has another distinct advantage, in that local authorities may award concessions to private companies to manage passive infrastructure on a utility basis. These concessions can be legal as long as 'passive infrastructure' is kept separate from the deployment and the operation of electronic communication networks and services. This type of concession is likely to enable public–private partnerships and allow the private sector to participate in the

Table 6.3 Legal situation of local authorities acting as telecommunications operators

Country	Identified legal barriers (other than competition law)	Maximum level of involvement: duct/dark fiber/optoelectronic equipment/full operation	Examples
Denmark	No legal barriers – but EU rules on state aid and competition law apply	No restrictions for local authorities on the maximum level of involvement. Third parties must be granted access to these infrastructures on fair and nondiscriminatory terms	None. Traditionally, local authorities have not been actively involved in telecommunications networks. There are a few cable TV and public utility companies – primarily electricity – controlled by or participated in by local authorities
Finland	None identified	In Finland, there are more than 50 local telecommunications networks and services operators. Traditionally, local municipalities have had significant interests in them. Many smaller local operators that are not quoted in the Stock Exchange are partly or fully owned by municipalities	According to FICORA, there is at least one example in the Helsinki area where the local electricity utility company has built up a fiber network. It is also understood that, at least, the City of Helsinki is planning to commence installing duct, tired of continuous digging of competing networks in the city
France	Article L.1511–6 of the Code général des collectivités territoriales. New roles for local authorities to be legalized by France's transposition of the e-commerce directive, i.e. La loi sur l'economie numerique, to be adopted this year by Parliament	Local authorities may establish infrastructures aimed at supporting telecommunications networks, but are not allowed to act as telecommunications operators. So, local authorities are allowed to lay dark fiber, but are not allowed to activate it; although this is subject to debate. The French authorities are working on the distinction between telecommunications equipment constituting passive infrastructures, and equipment necessary to activate a network	See CI study on Broadband stimulation in France, Ireland and Sweden (www.cullen-international.com select public area)

Continued

Table 6.3 Legal situation of local authorities acting as telecommunications operators (*Continued*)

Country	Identified legal barriers (other than competition law)	Maximum level of involvement: duct/dark fiber/optoelectronic equipment/full operation	Examples
Germany	Legal debate under way on to what degree local authorities can engage in commercial activities	See right-hand column	A large number of local authority-owned utilities, together with local public sector banks, have set up as city network operators (e.g. Netcologne in Cologne, ISIS in Düsseldorf, EWETel in Oldenburg). However, these operators are vertically integrated and provide primary infrastructure, telecom networks and services primarily to their own downstream operations
Ireland	None identified	Local authorities may be involved in full telecommunications projects, provided at least 60 % of the project is funded by the private sector. Otherwise, the project must be limited to passive infrastructure elements. However, third parties must be given open access to these infrastructures	See CI study on Broadband stimulation in France, Ireland and Sweden (www.cullen-international.com select public area)
Italy	None identified	—	There are several utility companies controlled or partly owned by local authorities. Metroweb, a dark fiber provider was owned 33 % by e-Biscom and 67 % by AEM S.P.A., a multi-utility company owned 51 % by the City of Milan. Atlanet, another licensed operator, is owned 33 % by FIAT Group. 33 % by Telefonica and 33 % by ACEA, a multi-utility company of the City of Rome

Continued

Table 6.3 Legal situation of local authorities acting as telecommunications operators (*Continued*)

Country	Identified legal barriers (other than competition law)	Maximum level of involvement: duct/dark fiber/optoelectronic equipment/full operation	Examples
Netherlands	None identified	Local authorities may be involved in development of telecommunications networks infrastructure with no restrictions on the maximum level of involvement. Third parties must be granted access to network infrastructure on fair and nondiscriminatory terms	Broadband access initiatives in the municipalities of Eindhoven and Helmond initiated by the Ministry of Transport, Public Works and Water Management in the framework of the Kenniswijk project. Fiber To The Home project initiated by the municipality of Almere in the framework of Almere Kennisstad project. The fiber network is intended as a community-owned network, fully unbundled and open for telecom operators and service providers
Spain	Local authorities can conclude agreements with telecom operators through 'mixed economy companies' or other types of formula	Telecommunications ventures must respect objective criteria and cannot discriminate between operators	One example is Metrocall, a company owned 40 % by the Metro of Madrid (Ayuntamiento, 75 % and Comunidad de Madrid 25 %) and 60 % by Euroinsta/Tecnocom S.A. It holds a type C1 license for the provision of support services to mobile operators in the Madrid metro network
Sweden	No legal barriers – but EU rules on state aid and competition law apply	Local authorities are generally not allowed to own the networks and act as telecommunications operators. They have to select the network installer and the network operator through competitive tenders. Companies owned by the local authorities are also eligible to participate in this procurement process. The successful operator is then required to provide open nondiscriminatory access to the network for other operators at fair prices. This operator would also become the owner of the network	See CI study on Broadband stimulation in France, Ireland and Sweden (www.cullen-international.com select public area)

Continued

Table 6.3 Legal situation of local authorities acting as telecommunications operators (*Continued*)

Country	Identified legal barriers (other than competition law)	Maximum level of involvement: duct/dark fiber/optoelectronic equipment/full operation	Examples
UK	There could be a problem for local authorities – while they have access to the highway, if they dig too far below the surface, they would need permission from the owners of properties that front the road, and definitely if they wanted to build up to the property	—	There are no examples of such infrastructure being provided or even being built. Oftel published guidance on its policy towards infrastructure sharing in June 2002. See www.oftel.gov.uk/publications/ind_guidelines/ duct0602.htm (see, in particular, Chapter 5 on municipal fiber networks). In Scotland, the Scottish Executive Agency has tendered for a contract to lay and manage a fiber optics infrastructure in 12 of the country's top business parks (Atlas project). The new capacity acquired will be offered as a neutral high-capacity link, which any telco can use

financing of this type of infrastructure in partnership with the public sector. In some cases, it has been argued that the creation of this type of 'civil infrastructure' by the public sector, or with public sector financial assistance, constitutes 'state aid' to those operators who lack physical infrastructure, and disadvantages operators who have already built their own infrastructure. This argument can be rebutted on the following grounds:

- Civil infrastructure is open to all operators in a truly nondiscriminatory manner. As a result, it does not create competitive advantages for any particular operator.
- Access is being provided at a fair price – neither for free, nor at a subsidized price.
- Under European Union law, state aid is illegal if it distorts competition. Building of civil infrastructure could help eliminate barriers and enhance competition in both communication networks and services, in a way that benefits both consumers and the overall economic welfare.

Interestingly, there is also case law confirming that the provision of infrastructure paid for out of public funds does not constitute aid, as long as it creates benefits for a wider range of users who pay for the usage of this infrastructure through direct or indirect charges.

Table 6.3 summarizes the legal situation pertaining to local authorities acting as telecommunication operators, in ten EU countries [1]. The table shows:

- whether any legal barriers for local authorities to engage in telecommunications ventures (other than competition law) have been identified;
- the maximum level of involvement of a local authority in a telecommunications venture;
- whether there are examples for municipal duct and dark fiber utilities, or even full telecommunications network operation, in the Member States.

6.6 Summary

In this chapter, first, the various models with different levels and nature of public entities' participation were analyzed. We can conclude that, most likely, there is no 'one fits all' type of solution. Each community or local government has to choose what would suit best its own needs and requirements. The size of the community, the type and number of existing businesses located in each community, possible legal restrictions, the needs of the public sector, the potential of the community to attract economic investment, as well as the aspirations of the community for the future, are some of the elements that will determine which model or type of intervention is most suitable. And no comparison is void of subjective elements.

Since, in several cases, one might wish to combine some of the models, we also discussed which combinations are potentially good ones, which are possible, and which models cannot be combined because of competition issues.

Finally, our aim with the overview of the legal environment was to provide some starting points for public entities in order enable them to identify issues where it is necessary to seek specific legal advice.

Acknowledgments

The author wishes to acknowledge the following for their invaluable contributions to the material in this chapter, which was first published by the European Regional Information Society Association (eris@) as *A Guide to Regional Broadband Deployment*: Ilkka Rasanen (Intel), Dan Kiernan (Alcatel), Meni Styliadou (Corning), Christian Ollivry (Motorola), Maurice Sanciaume (Agilent) and Francisco Fuentes (Cisco).

Reference

1. eris@ (2003), *A Guide to Regional Broadband Deployment*, 1st edition, Brussels.

7

Customer-Owned and Municipal Fiber Networks

Bill St. Arnaud
CANARIE Inc., Ottawa, Canada

7.1 Introduction

During the last couple of years, a revolution has been taking place in high-speed network infrastructures around the world, but particularly in Canada, and more especially in Quebec. At root, this revolution is driven by the availability of low-cost fiber optic cabling. In turn, lower prices for fiber are leading to a shift from carrier-owned infrastructure toward more customer- or municipally owned fiber, as well as to innovative sharing arrangements, such as fiber condominiums.

'Condominium' fiber is installed by a private contractor on behalf of a consortium of customers, with customers owning the individual fiber strands. Customers/owners light their fibers using their own technology, thereby deploying a private network to wherever the fiber reaches, perhaps including carrier Central Offices and Internet providers. The business arrangement is comparable to a condominium apartment building, where common expenses, such as management and maintenance fees, are the joint responsibility of all the owners of the individual fibers.

'Municipal' fiber is owned by a municipality (or perhaps a public utility), with its key property being that it has been installed as public infrastructure, with the intention of leasing it to potential users or network builders. Again, lighting the fiber to deploy private network connections is the responsibility of the lessee, not the municipality.

7.1.1 Customer-owned dark fiber

With customer-owned dark fiber networks, the end customer owns and controls the actual fiber and decides to which service provider they wish to connect for services such as telephony,

Broadband Services: Business Models and Technologies for Community Networks. Edited by I. Chlamtac, A. Gumaste and C. Szabó © 2005 John Wiley & Sons, Ltd. ISBN 0-470-02248-5.

cable TV and Internet. However, most customer-owned dark fiber deployments are used for delivery of an Internet service.

Professional 3rd party companies, who specialize in dark fiber systems, take care of the actual installation of the fiber, and also maintain it on behalf of the customer. Technically, these companies actually own the fiber, but sell IRUs (Indefeasible Rights of Use) for up to 20 years for unrestricted use of the fiber. These companies are also responsible for managing the complex relationship of agreements with the municipality and others on use of rights of way, construction, maintenance, etc.

There is no additional management complexity or overhead associated with customer-owned dark fiber. In fact, in many cases, customer-owned dark fiber may be more reliable than traditional telecommunication services, and easier to manage because it vastly simplifies the network architecture and allows the consolidation of network services to a central hub.

In addition, customer-owned dark fiber provides for increased competition among service providers and levels the playing field amongst all service providers for the delivery of telecommunication services. With customer-owned dark fiber, customers build networks to carriers, rather than the traditional model where carriers build networks to customers.

With customer-owned dark fiber, the customer can now interconnect to the carrier(s) of their choice at a convenient meet-me point, such as a school board office, municipal building or a carrier neutral collocation facility. This is particularly advantageous for the interconnection to new smaller innovative communication companies or ISPs, who cannot afford to build expensive physical network infrastructure.

In essence, customer-owned fiber is moving the demarcation point at which the carrier interconnects to the customer. In the old days of telephone monopoly, the demarcation point was in the customer's home, as the carrier owned the telephone and the inside wiring. With competition, the demarcation point has moved to the edge of the customer premises. But with low-cost fiber and new LAN-based Ethernet technologies, the demarcation point is moving closer and closer to the carrier.

7.1.2 Condominium fiber

All across North America, businesses, school boards and municipalities are banding together to negotiate deals to purchase customer-owned dark fiber. A number of next generation service providers are now installing fiber networks and will sell strands of fiber to any organization that wishes to purchase and manage its own dark fiber.

Many of these new fiber networks are built along the same model as a condominium apartment building. The contractor advertises the fact that they intend to build a condominium fiber network, and offers early participants special pricing before the construction begins. That way, the contractor is able to guarantee early financing for the project and demonstrate to bankers and other investors that there are some committed customers to the project.

The condominium fiber is operated like a condominium apartment building. The individual owners of fiber strands can do whatever they want with their individual fiber strands. They are free to carry any type of traffic and terminate the fiber any way they so choose. The company that installs the fiber network is responsible for overall maintenance and repairing the fiber in case of breaks, moves, addition or changes. The 'condominium manager' charges the owners

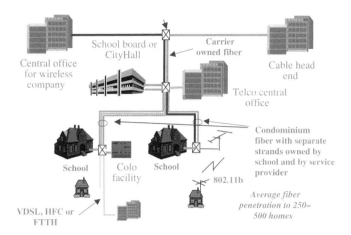

Figure 7.1   Municipal/condo architecture

of the individual strands of fiber a small annual maintenance fee, which covers all maintenance and right of way costs.

The architecture of a municipal/condominium fiber network is illustrated in Figure 7.1.

7.1.3 Community or municipal fiber networks

Many municipalities are now looking at the advantages of condominium or customer-owned fiber. Not only do these networks substantially reduce telecommunication costs, but they can also significantly increase the number of competitive service offerings, because any service provider can also purchase strands of fiber.

It is not necessary for the municipality or community to own the fiber or build the networks themselves. Municipalities and communities can encourage the deployment of condominium fiber networks in their jurisdiction by tendering their existing telecommunication business only to those companies that will deploy such networks. In some cases, provincial, state and federal governments can play a critical leadership role by providing additional funding to make sure that all communities can enjoy the benefits of condominium fiber networks.

7.2 Advantages of customer-owned dark fiber/municipal fiber to different users and businesses

7.2.1 Public institutions

The organizations that are usually the earliest beneficiaries of dark fiber are schools and school boards. The initial primary driver for dark fiber by individual customers is the dramatic savings in telecommunication costs. The reduction in telecommunication costs can be in excess of 1000 %, depending on bandwidth requirements.

A typical dark fiber connection may cost a one time fee of $25 000 (2003 figure). If an organization is currently leasing an OC-3 circuit (155 Mbps), it could be paying anywhere between \$3000 – \$6000 per month, which results in an annual cost greater than purchasing dark fiber. When leasing local loops with greater capacity than OC-3, the cost savings can be more dramatic.

The typical pay-back for dark fiber, as opposed to purchasing managed bandwidth, is 12 to 18 months. And for this short pay-back, the customer gets a 'future-proof' network for the next 20 years, where there is no increase in local loop costs as the customer's bandwidth demands increase, except to upgrade the equipment at the ends of the fiber.

Some of the benefits of customer-owned dark fiber include:

1. Significantly reduced local loop costs for connecting each school to the central school board office. The typical average cost to connect up a school is \$US 25 000 per school for a 20-year IRU. In some cases, the fiber can be leased at a price of typically around \$400 per month per school.
2. Reduced network management complexity, in that only simple and easy-to-use Fast Ethernet or Gigabit Ethernet transceivers are needed at each end of the fiber.
3. No additional costs to increase the bandwidth of the network, other than to upgrade the transceivers at the end of the fiber.
4. A dramatic reduction in the number of network Web and LAN servers. With dark fiber, the individual school LANs can be extended to the central school board office. All of the servers can be relocated to the central site and aggregated into single-server systems. This significantly reduces network management and complexity.
5. A significant reduction in network staff and travel, as most of the LAN and network servers are relocated to a central site.
6. A greater choice of service providers for Internet and other advanced services. Those school boards that use a carrier to connect to the school are limited to, in most cases, one or two carriers who can connect up the schools in their district. But if the schools purchase their own fiber, they can aggregate their traffic and then connect to a greater choice of service providers at the central school board office. It also makes their network business more attractive for outsourcing.
7. Development and deployment of new applications and services that would not be possible with a limited bandwidth service. Many schools with dark fiber, for example, are putting Internet telephones on each teacher's desk. With a dark fiber network, there is no additional cost to provide such a service, except for the purchase of the telephones themselves. Other applications include videoconferencing and distance education.

However, school boards should be careful to compare total solutions packages, where a carrier owns and manages the fiber to the school, versus purchasing their own dark fiber. Many school boards and governments are enticed with the worry-free management of a wide area network, where the carrier owns the fiber to the school or other facility. While this may be attractive at first, it will limit the school in terms of its future options. When a school board goes to renew its contract a few years down the road, the incumbent carrier who installed the initial fiber will have a competitive advantage over any other new entrant in the marketplace who must now also install fiber to the school.

As most schools are located in suburban/rural areas, there is rarely a business case for a second carrier (never mind a third or fourth) to build a fiber network to the school. When the school comes to renew its service contract, it will discover that the original provider will have a significant competitive advantage over any other service provider.

Schools, because of their geographic location and because they are usually the first organization to acquire dark fiber in the community, have a responsibility, perhaps more so than any other organization in the community, to ensure a level playing field and not to confer competitive advantage to any single carrier.

In addition to the many benefits of dark fiber for schools and large businesses, as listed above, customer-owned dark fiber has a number of significant advantages for universities and research institutions:

1. It significantly reduces telecom costs in connecting satellite campuses.
2. It allows the deployment of new high-bandwidth applications without a bandwidth cost penalty.
3. It allows the testing and deployment of new optical technologies that would not be possible with a managed carrier service.

7.2.2 Business organizations

Many large businesses have been acquiring dark fiber for some time now. The advantages for a large business are as follows:

1. A significant reduction in local loop telecom costs.
2. Centralization of servers from remote offices scattered around a city.
3. The ability to capitalize telecom expenses, rather than treating them as an ongoing service cost.
4. The ability to deploy redundant paths to multiple carriers. Usually, in the past, this has been done by purchasing separate SONET services from different carriers at considerable cost.
5. Outsourcing LAN, storage and network servers.
6. Relocation of speed-sensitive network servers to a server farm. Normally, speed-sensitive servers, such as LAN-based video and audio servers, have to be located on the LAN next to the user. However, as there are no speed limitations with dark fiber, these servers can be easily located at a central server.

A number of integrators and outsourcing companies will arrange the dark fiber connections for a large business as part of the outsourcing package.

Surprisingly, even small businesses with fewer than 30 employees can benefit from customer-owned dark fiber. This is particularly true of new high-tech e-commerce, or other web-based businesses. The advantages for a small business are as follows:

1. A significant reduction in local loop telecom costs.
2. The ability to capitalize telecom expenses, rather than treating them as an ongoing service cost.
3. 'Customer facing' servers, such as e-commerce servers, web hosting servers, etc., can be located at a carrier neutral collocation facility, where there is redundant power, 24-hour security, and multihoming services. If the telecom link between the collocation facility and the office goes down, no customers are affected.

4. Outsourcing LAN, storage and network servers.
5. Relocation of speed-sensitive network servers to a server farm.

Many service providers cannot afford to build a large telecommunications infrastructure. Carriers who have built such infrastructures under government regulated monopoly have a significant advantage in the marketplace. However, if the customer owns the fiber, then they can connect to the service provider of their choice at a carrier neutral collocation facility. Customer-owned dark fiber levels the playing field and gives all ISPs an equal opportunity to capture the customer's business.

The customer can delegate management of the fiber to the service provider of their choice for the life of the service contract. The service provider is then responsible for fiber maintenance and end-to-end service. However, when the customer owns the fiber, they can change service providers at any time (or at the end of the service contract) to some other service provider, and enter into a similar arrangement, where the new service provider is responsible for end-to-end performance.

Carrier neutral collocation facilities are an essential component of providing such an arrangement. Carrier neutral collocation facilities with fiber 'meet-me' rooms, allow the organizations with customer-owned fiber to easily switch from one service provider to another.

7.2.3 Office building owners and managers

Increasingly, large building owners are facing congestion in their telecommunication risers from all the new entrant telecom companies who want access to their tenants. A number of large building owners have now declared a moratorium on new entrants. In fact, some building owners are proposing that no carriers be allowed into the risers. Instead, the building owners themselves are installing fiber in the risers (usually two risers for redundancy purposes) from each tenant to an equipment room in the basement, where they interconnect to the carriers.

But even that is only a partial solution. The access ducts to the building are becoming congested, as well as the equipment room in the basement. The next logical step is to extend the tenants' fiber beyond the basement to a nearby open collocation facility. The building owner may acquire 48 or more strands of customer-owned fiber to at least two separate collocation facilities. As part of the leasehold package, the tenant can lease this fiber for the duration of their building lease to access a service provider of their choice at the collocation facility.

In many cities, companies are building collocation facilities to allow the interconnection of networks between competing service providers and for the hosting of web servers, storage devices, etc. In fact, many of these facilities are being built by the same companies that also own large high-rise buildings. The collocation facilities are rapidly becoming the obvious location for terminating customer-owned dark fiber. With a simple change in the optical patch panel in the collocation facility, the customer can quickly and easily change service providers at very short notice.

The tenants in such buildings can then lease several strands of dark fiber within these trunks to one or more of the carrier neutral collocation facilities.

By leasing fiber to separate collocation facilities, the tenant is assured of route diversity and redundancy in case of a fiber break. The building tenants can now easily outsource their web and network services to a number of competitive outsourcing companies located at the carrier neutral collocation facility. With dark fiber, there are no worries about congestion or bandwidth bottlenecks for the business when accessing its own server.

In summary, the benefits of customer-owned dark fiber to large building owners are:

1. It eliminates congestion in the risers, and thereby saves valuable floor space.
2. It significantly reduces the cost of telecom services for tenants.
3. It reduces congestion in the building access ducts and machine rooms.
4. It allows for a new line of business to give the tenants the ability to outsource their server to a collocation facility, particularly if that facility is also owned by the building owner.
5. It provides for a greater choice of server providers for the tenants, in being able to access a greater number of service providers at the collocation facility.
6. It allows the tenants to establish redundant and diverse paths to two different service providers at two different collocation facilities.

7.2.4 Consumers or home owners

The fiber-to-the-home (FTTH) market is still in its infancy. However, customer-owned dark fiber is increasingly being seen as a superior alternative to the current proposed approaches for FTTH by the telephone and cable companies.

A number of entrepreneurs and small companies are exploring concepts where the home-owner has title to their own strands of fiber in a neighborhood FTTH build. The advantage of this approach is that the FTTH systems can be deployed in areas of low density and very small take-up (i.e. less than 5 %). Rather than treating the network as an expense, the network infrastructure is seen as an investment by the homeowner. When an FTTH system is first deployed, the initial customers (individual homeowners or other parties) make an investment of between $10 000 and $40 000 for each customer drop. Their investment is secured by IRUs on a number of fiber strands – anywhere from 12 to 96 strands. As additional customers in the neighborhood are signed up for the service by the condominium fiber manager, the strands are purchased by the new customers from the initial investors. This way, the initial investors can see a significant return on their investment (as much as 500 % return) and yet have the confidence in knowing that their investment is secured by the underlying IRUs (as opposed to shares in a CLEC or condominium fiber company).

Several companies are now developing technology for delivery of Ethernet to the home. The beauty of this technology is that it uses a well-known open standard for the delivery of service. The technology can be used to support voice, video and extreme high-speed data.

Ethernet delivery of services can be best accomplished with optical fiber. The big challenge for governments is whether the deployment of broadband services to the home should be a monopoly service, like the cable and telephone model, where the service provider owns the infrastructure, or an open model, where the consumer owns (and delegates management to a service provider of their choice) the infrastructure and connects to the service provider of their choice at an open collocation facility.

Carriers who have built telecom infrastructures under government regulated monopoly have a significant advantage in the marketplace. If consumers own or control the infrastructure, they can then connect to the service provider of their choice. More importantly, it levels the playing field, as it allows small service providers who cannot afford to build a costly infrastructure to sell services to the consumer.

7.2.5 Municipalities and governments

When independent carriers deploy their own fiber networks to the customer premises, each provider needs to deploy separate fiber cable to virtually every building and within every riser in each building. This results in a significant number of independent fiber cables, and hence road trenching, required to interconnect all the downtown buildings with the multitude of service providers.

The other result is that many municipalities are declaring moratoria on digging up the roads by carriers who want to install new fiber cables. In addition, many cities are refusing to grant new construction permits for any road section that has been repaved or rebuilt within the last five years. The companies that have already deployed fiber then have a de facto monopoly for the next five years. If customer or service providers cannot purchase dark fiber at a reasonable cost, it then becomes very difficult for downtown businesses to connect to competitive service providers.

In recognition of this issue, some cities are being proactive and are providing incentives or mandating that carriers install additional open access conduit when they dig up the road to install new fiber, or provide open access fiber at an agreed upon price. Municipalities can leverage their existing telecommunication procurements and their private rights of way to negotiate open access fiber builds throughout their jurisdiction.

An excellent example of this approach is the Chicago CivicNet project, where the City of Chicago offered its annual telecommunications budget of $US 25 million and access to city ducts, subway tunnels and other facilities to companies who would install open access condominium fiber to all public sector buildings in the city. For more information, see [1].

The province of Alberta in Canada has launched a similar initiative for all 430-plus municipalities in that province, in a project called SuperNet [2]. In this example, the province offered its telecommunication business plus $CDN 183 million to a consortium of carriers to deploy open access condominium fiber to all schools, libraries and hospitals in every community in the province. The province holds title to the IRUs on all the fiber and then earns back its investment as the fiber strands are sold off to competitive service providers, school boards, etc., see [3].

7.3 Deployment of municipal fiber networks

7.3.1 Rights of way and Municipal Access Agreements (MAAs)

Deploying fiber networks involves entering into legal contracts with both the local municipality and owners of any existing support structures, such as poles and ducts. The public right of way is a crowded and complex environment to work in, whether on surface, subsurface or aerial infrastructure. Unfortunately, there is far, far more involved than simply tacking some cable onto existing utility poles, albeit the use of existing poles would appear to be the least complex approach for a fiber builder.

By law, if someone places cable in or on someone else's support structures in the public right of way, two legal rights are involved. First, it is necessary to have the permission of the pole (or duct) owner to install the cable. This involves the use of someone else's property and, as with any other commercial transaction, the owner of the property has the right to charge compensation in accordance with policies established by national and local regulators. The usual process is to enter into a Support Structure Agreement (SSA) with the pole or duct owner.

The pole or duct owner may have their own set of requirements, including a number of upfront and ongoing insurance and liability requirements. In addition, records and drawings must be professionally done and provided to the support structure owner.

The second right involved is the right to use the public right of way, i.e. municipal property. Most cities require any organization that is deploying a fiber network on PROW to enter into a Municipal Access Agreement (MAA). The pole (or duct) owner would have been given (in his MAA) a very specific right to use the public right of way for a very specific purpose (like an easement). The pole owner has no legal right to pass on the right to use the public right of way to any other party who may wish to place cable on the owner's poles. Thus, any other entity that wishes to place cable on the owner's poles, in addition to an agreement with the pole owner, requires its own MAA with the municipality for the use of municipal property. This MAA would provide the necessary indemnification and all the other protection and assurances that the municipality, on behalf of the public, needs.

As part of the MAA, there are usually a number of upfront and ongoing insurance and liability requirements. Records and drawings must be professionally done and provided to the municipality. The fiber builder is usually responsible for any future relocation costs, should that be necessary.

Compensation to the municipality is usually required. Most municipalities' policy is to recover all costs from proponents to ensure that any costs incurred by the municipality are ultimately borne by those who specifically benefit, and not by the general taxpayer. The property taxpayer should not be progressively out-of-pocket as a result of an increasing use of the public rights of way by private entities. Such a back-handed subsidy might also induce certain uses to proceed when they would not otherwise be viable without specific public support. Such a situation might also inhibit the evolution and deployment of more efficient and effective uses. Some municipalities also charge a fee for the use of municipal property.

7.3.2 Carrier neutral collocation facilities

In many cities, companies are building facilities to allow the interconnection of networks between competing service providers and for the hosting of web servers, storage devices, etc. They are rapidly becoming the obvious location for terminating customer-owned dark fiber.

These facilities feature diesel-powered backup systems and the most stringent security systems. The facilities are open to carriers, web hosting firms and application service firms, Internet service providers, etc.

Most carrier neutral open collocation facilities feature a 'meet-me' room, where fiber cables can be cross-connected to any service provider within the building. With a simple change in the optical patch panel in the collocation facility, the customer can quickly and easily change service providers at very short notice.

Many of these concepts of carrier neutral collocation facilities were first developed with the next generation Internet programs in the United States and Canada, with a concept called a GigaPOP. Leading researchers and universities recognized that there were many benefits to interconnecting to carriers at a common 'meet-me' point. So, rather than having multiple carriers build separate facilities to university campuses, the universities instead built one single telecommunication facility to a GigaPOP and then interconnected to one or more carriers on a new demarcation point that was not on the customer premises.

When selecting a fiber provider, care should be taken to see if their fibers terminate at carrier neutral collocation facilities. Some fiber providers only terminate their fiber in their own central offices, which makes it difficult to interconnect to other service providers or attach your own equipment to the fiber.

7.3.3 Equipment to light up dark fiber

With customer-owned dark fiber, simple laser devices called transceivers are all that is required to light up the fiber. These devices will work with SONET, ATM and Ethernet devices at either end of the fiber connection. As such, there are only three things that can go wrong with a customer-owned dark fiber – the source transceiver, the destination transceiver or the fiber itself.

Transceivers for Ethernet data can drive fiber up to 120 km. The following are typical distances and prices for Ethernet transceivers:

1. Fast Ethernet (100 Mbps) transceivers can drive fiber up to 80 km and will cost about $US 700 per end.
2. Gigabit Ethernet transceivers will drive fiber up to 60 km and will cost about $US 2000 per end.

Prices for transceivers are dropping dramatically. Already, 10 Gigabit Ethernet transceiver chip sets, that will drive 40 km of fiber, are being sampled at less than $US 100. These transceivers usually can be controlled and managed by standard LAN management systems.

7.4 Cost analysis

7.4.1 Overview

Customer-owned dark fiber is still a very immature industry. As such, costs for fiber can be extremely variable, from as little as $0.50 per meter per strand, to $6.00 per meter per strand, for a 20-year IRU. Part of the problem is simple economies of scale. The biggest cost component of fiber networks is the installation cost. The installation cost is virtually the same, whether you are installing a 12- or 864-strand cable. Fiber networks deployed with 12 to 48 strands have a much higher cost per strand than 864-strand systems. Other complicating factors depend on whether roads have to be trenched to lay down the fiber, or whether existing support structures, such as poles and conduit, can be used for the installation of the fiber.

For budgetary purposes, $2 – $3 per meter per strand-pair can be used for a 20-year IRU for existing fiber. Additional strands do not significantly increase the cost. So a budgetary price for four or six strands of fiber may be around $4 per meter.

The key thing to note is that this is the one time upfront cost for the purchase of a 20-year IRU. The IRU can usually be considered as a physical asset which can be resold, traded or used as collateral. As such, the cost of an IRU can be amortized over its 20-year lifetime, which results in a monthly cost substantially below traditional telecommunication services.

Ultimately, as the industry matures, and large optical cables with 864 strands or greater are routinely deployed, the cost of dark fiber is expected to drop down to $0.07 to $0.10 per meter per strand.

7.4.2 Some detailed cost components of customer-owned dark fiber

Dark fiber is made up of four different cost components:

1. Trunks.
2. Laterals.
3. Building entrances.
4. Termination panels.

Trunks are the main fiber cables that may carry hundreds of fiber strands owned by a variety of carriers and institutions. Laterals are the fiber cables from the customer premises to the nearest splice point on the cable trunk. Generally, laterals are used exclusively by the customer and therefore the customer must pay for the full cost of the cable and its installation. Within cities, laterals can be as short as a few meters. They can extend several kilometers in suburban and rural areas.

In some cases, the costs of a lateral, particularly in suburban and rural areas, can be more costly than the much longer fiber run on a trunk cable.

The minimum size of a lateral is usually 12 strands. But even though a lateral may have 12 strands, only two or four of those strands may be spliced to dedicated fibers on the trunk. Most fiber-provisioning companies provide additional spare strands on the trunk, to which the customer can connect at a later date for an additional cost.

Building entrances and termination panels are the facilities within the customer's premises for the termination of the fiber. As a rule of thumb, a building entrance with fiber panel termination is about $US 5000, but may vary from as little as a few hundred dollars to $15 000 – $20 000. The large variation in cost is due to many factors, including whether the installers have to drill through concrete walls to terminate the fiber and/or bring up the fiber several floors in a riser.

For either trunks or laterals, the basic cost calculation is the same. Overwhelmingly, the single biggest cost is the installation of the fiber itself. On fiber trunks, the cost of the installation is shared amongst the owners of the individual fiber strands, and so, on a large cable trunk of 864 fibers, the installation cost per strand can be quite small. On laterals, there is generally no other user, so the customer must pick up 100 % of the cost.

There are four types of fiber installation:

1. Aerial on existing poles ($3 – $6/meter).
2. Buried in existing conduit ($7 – $10/meter).
3. Jet fiber in micro conduit ($3 – $15/meter).
4. New trenching and laying of conduit ($35 – $200/meter).

Installation charges are almost entirely made up of labor costs, so the numbers quoted above are pretty well the same, whether they are stated in Canadian or US dollars.

Aerial installation on existing poles is by far the cheapest installation method and the most reliable. Most regulatory bodies have well-established rules and procedures for licensed carriers and fiber installers to access existing utility and telephone poles. The variation in the cost of aerial installation is largely dependent on how accessible the poles are from the street. If the poles are in backyards and only accessible by foot, then the installation costs are at the higher end of the range stated above. If the poles run right along the roadside, then installation can be done directly from a truck on the roadside.

Installation in existing conduit is the next best option. Many municipalities and regulators require carriers to install extra conduit accessible by any other licensed carrier or fiber installer. As with poles, regulators have set prices for the cost of access to this conduit.

Jet fiber is a new approach to fiber deployment. In this case, the fiber provider only installs 'microconduit' instead of fiber. When a customer requires a fiber pair, it is blown into the microconduit on demand. The advantage of this approach is that far fewer splices are required and the fiber can be blown all the way into the customer's premises. Currently, the capital cost of jet fiber is higher than for traditional approaches, but it is expected that the ongoing systems costs will be lower.

If there are no existing conduits or poles, commonly referred to as support structures, then a fiber trench must be dug and new conduit installed. This is by far the most expensive approach. In the downtown, core trenching costs can be prohibitive because of the obvious disruption of traffic and the complex existing ductwork that already lies beneath most of our downtown streets.

A new approach that is coming onto the market is installing fiber in sewer lines. This technology appears to be 1/3 to $^{1}/_{2}$ the cost of new trenching.

There are a number of different approaches if you are forced to deploy new fiber where there are no existing support structures, such as poles and conduits:

1. Traditional trenching and conduit deployment with a backhoe.
2. Direct bury with a 'fiber plough' where no conduit is deployed.
3. 'In the groove' technology, where a very narrow groove is cut into the existing roadbed.
4. 'Sewer' systems, where robotic systems install fiber in storm or sanitary sewers using either specialized cable or stainless steel tubing.
5. 'Gas' pipeline systems, where the fiber is installed in active gas pipelines.
6. 'Directional boring,' where, for short distances, a tunnel can be bored laterally underneath the ground.

In any event, trenching of any kind should be avoided as much as possible. Fiber on poles in most situations results in the most reliable installation, even over buried fiber.

When planning to build a condominium fiber network, it may be necessary to add an additional $1 – $3/meter for engineering, design, supervision and installation. In addition, condominium cables are priced on a section by section basis, with the price varying depending on how many users are in a section and the specific costs of each section.

The cost of fiber cable itself varies depending on the number of strands in the fiber cable and whether it is a standard single mode fiber or a specialized low-attenuation type of fiber.

The following costs are typical costs of SMF-28 fiber cable in US dollars:

- $0.15 per strand per meter for 36 strands or less;
- $0.12 per strand per meter for 96 strands or less;
- $0.10 per strand per meter for 192 strands or less;
- $0.05 per strand per meter for more than 192 strands.

The standard minimum cable size is 12 strands. So the lowest cost for a lateral cable is $1.80 per meter for the cable, plus $3 – $6 per meter for the installation. As a rule of thumb, laterals are priced at $5 per meter aerial, $10 per meter in existing conduit and $35 per meter if new trenching is required.

On large cable fiber trunks, a budgetary number of $0.50 per meter per strand (i.e. $1 per meter per pair) is reasonable. The cost of installation is negligible on a per strand basis.

7.4.3 Ongoing yearly costs for customer-owned dark fiber

As a rule of thumb, the annual maintenance and right of way charges for dark fiber amount to 5 % of the capital cost.

In some cases, fiber installation companies waive all annual maintenance costs for schools and hospitals if other strands of fiber in the cable bundle are used to carry commercial services.

There are two components to the annual charges:

1. Right of way charges.
2. Annual maintenance charges.

Both charges are assessed against the fiber cable and not the individual strands. So, once again, the per strand cost of these charges can be very small for large strand cables.

Government regulators, such as the FCC in the United States and the CRTC in Canada, have established set prices for the costs of rights of way on public land, or for regulated telecommunication facilities. In the United States, the situation can become more complex because Public Utility Commissions may have a different set of rules than the FCC.

The regulated telecommunication facilities are called support structures and are devices that are currently used to carry regulated telecommunication services. So, generally, existing telephone poles, utility poles and telephone company conduit is covered under these regulations. However, utility poles that do not carry any existing telecommunication facilities, railroad bridges and road bridges are generally not covered by these regulations. The owners of these facilities do not have to provide access to other carriers to their facilities, and if they do, they can charge any price the market will bear.

As a rule of thumb, the following are the typical right of way costs:

- $1 per pole per month;
- $0.50 per strand of support wire between poles per month;
- $1 per meter of existing conduit.

Right of way costs therefore work out to be typically $0.50 per meter per year on poles and $1 per meter per year in an existing conduit. Many cities are also assessing right of way charges in the downtown core to reflect increased costs to the city for traffic disruption, etc. These costs typically average $20 per meter per year.

It is important to note that these costs are assessed against the fiber cable. The more fibers in the cable, the lower the cost per strand.

Maintenance charges vary between $150 to $250 per kilometer (or mile) per year.

7.5 Operation and management

7.5.1 Reliability of dark fiber

Customer-owned dark fiber can be more reliable than traditional telecommunication services, particularly if the customer deploys a diverse or redundant dark fiber route.

Dark fiber is a very simple technology. It is often referred to as being technologically neutral. Sections of dark fiber can be fused together, so that one continuous strand exists between the customer and the ultimate destination. As such, the big advantage of customer-owned dark fiber is that no active devices are required in the fiber path. Since there are no active devices, customer-owned dark fiber, in many cases, can be more reliable than a traditional managed service. Traditional managed services usually have a myriad of devices in the network path, such as SONET multiplexers, add/drop multiplexers, ATM switches, routers, etc. Each one of these devices is susceptible to failure, and that is why traditional carriers have to deploy complex networks and systems to ensure reliability and redundancy.

Many customers assume that, because the carriers deploy SONET rings, they have a reliable network. In fact, SONET rings are generally only deployed between the carrier's central offices. Most customers today, except in exceptional circumstances, only have one unprotected link to the nearest central office, and this is the single weakest link in their network.

For the greatest reliability, many customers will install two separate dark fiber links to two separate service providers. Even with the additional fiber for redundancy, customer-owned dark fiber networks are cheaper than managed services from a carrier.

With dark fiber, customers have a number of choices in terms of reliability and redundancy:

1. They can have a single unprotected fiber link and have the same reliability as exists today with their current carrier.
2. They can use alternative technology, such as a wireless link for backup in case of a fiber break.
3. They can install a second geographically diverse dark fiber link, whose total cost is still cheaper than a managed service from a carrier.

Because fiber has a greater tensile strength than copper or even steel, it is less susceptible to breaks from wind or snow loads.

7.5.2 Operation and maintenance

In most cases, management of the fiber is contracted out to a 3rd party who specializes in the repair and maintenance of fiber networks. In many cases, these are the same companies who maintain and repair for the major carriers. They offer the same terms and conditions to dark fiber customers as they do for the major carriers.

Frequently, the companies that installed the fiber are also the ones who maintain the fiber. These companies will also look after any ongoing moves, additions and changes, as well as relocating the fiber in case of road construction and so forth.

Moves, additions and changes are generally quite trivial, and can be carried out by the fiber maintenance company on a routine basis. Most moves, additions and changes only require breaking and fusing together existing fiber pairs. The work can be done on an hourly rate, or priced on a per move/addition/change basis. The cost, terms and conditions for moves, addition and changes are usually included in your fiber maintenance agreement.

On very rare occasions, fiber has to be relocated because of road construction or repair. Usually, the city that has undertaken the road work will pay for the majority of the fiber relocation costs. However, if this not the case, minor relocations of several hundred meters are usually included as part of the maintenance contract.

If a major relocation is required, most contracts give the customer the option of paying their share of the relocation expenses in proportion to the number of fibers they own in the cable, or canceling the fiber ownership contract.

Although uncompensated fiber relocations are very rare, it is a factor that should be taken into account when purchasing dark fiber.

7.6 Summary

In this chapter, we focused on planning, implementation and operational issues of dark fiber networks. Dark fiber is considered to be technologically neutral and, in many cases, presents the least problematic construction from a legal/regulatory point of view. We showed that customer-owned dark fiber can be the right solution for school boards, municipalities and other public institutions, as well as for businesses. Different forms of ownership, operation and management of dark fiber networks were discussed. Finally, our cost analysis showed the attractiveness of dark fiber from a financial point of view.

References

1. http://www.cityofchicago.org/CivicNet/civicnetRFI.pdf
2. www.albertasupernet.ca
3. http://www.canarie.ca/advnet/workshop_2000/presentations/cheney.ppt.

8

Towards a Technologically and Competitively Neutral Fiber-to-the-Home (FTTH) Infrastructure

Anupam Banerjee and Marvin Sirbu

Carnegie Mellon University, Pittsburgh, USA

8.1 Introduction

The (hitherto) lack of initiative among incumbent local exchange carriers has forced local governments and communities to take interest in fiber-to-the-home (FTTH)[1]. Today, many such communities are making fundamental choices of technology and architecture, designing networks, planning deployment strategies and determining the range of services to offer. Municipalities and community associations are likely to have a greater interest (in economic welfare, and hence) in competition. Against this industry backdrop, in this chapter, we consider the implications of engineering, architecture, economics and ownership for competition in the FTTH industry.

We first discuss what we mean by competition in the telecommunications industry (Section 8.2). We then consider the engineering economics of four different FTTH network architectures (Sections 8.3 and 8.4): (i) home run fiber; (ii) active star; (iii) passive star (or passive optical networks – PONs); and (iv) wavelength division multiplexed passive optical networks (WDM PONs). Further, we define different models for competition in the FTTH industry. Results from the engineering cost models of these architectures in three different deployment scenarios:

---

[1] The FCC triennial review seems to have created a lot of interest in FTTH among the ILECs, but it remains to be seen if this interest will translate into initiatives in the near future.

Broadband Services: Business Models and Technologies for Community Networks. Edited by I. Chlamtac, A. Gumaste and C. Szabó © 2005 John Wiley & Sons, Ltd. ISBN 0-470-02248-5.

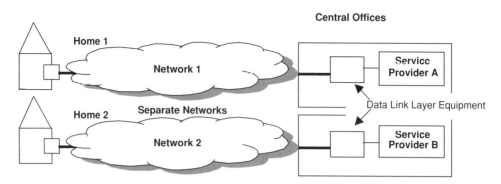

Figure 8.1 Facilities-based competition

(i) urban; (ii) suburban; (iii) rural are then used to comment on the implications that network architecture has for competition (Section 8.5). We show that the lowest cost FTTH architecture supports different models of FTTH competition and conclude with a discussion on issues in the FTTH industry structure (Section 8.5).

8.2 Models for competition in telecommunications

Competition in the telecommunications services industry can be facilities-based or non-facilities based; the Telecommunications Act of 1996 contemplates both forms of competition.

8.2.1 Facilities-based competition

Under this arrangement, each service provider serves the market using its own physical network (Figure 8.1). In the United States, the most common example of facilities-based competition is the mobile personal communications services market, where each mobile telephony services provider builds, owns and maintains its network.[2]

8.2.2 Non facilities-based competition or service level competition

In this context, each service provider shares the resources of a common network to provide service to its customers. We consider two models for this sharing.

8.2.2.1 Unbundled network elements (UNE)-based model for competition

Each service provider colocates its data link layer equipment at the CO and offers voice, data, video and data link layer services to its customers by renting 'unbundled network elements' (like a copper loop) from the network owner (Figure 8.2). The local telephone service industry exhibits this model of competition with CLECs (competitive local exchange carriers) renting UNE-loops from the incumbents to provide telephone or DSL services. For UNE-based[3] competition to be possible, physical plant unbundling must be feasible.

---

[2] However, in order to reduce costs for 3G wireless deployments, European mobile operators are moving to shared-cell infrastructures.

[3] Henceforth, by UNE we allude to UNE-loops.

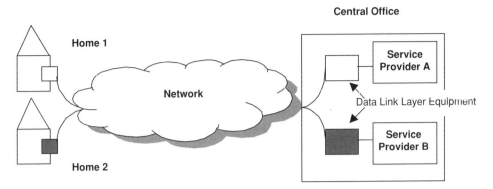

Figure 8.2   UNE-based competition

8.2.2.2 Open access-based model for competition

Each service provider has to share the common data link layer (generally belonging to the network owner) in order to provide voice, video or data services (Figure 8.3). Multiple ISPs providing Internet services over a single cable TV network would be an example.

8.3 Fiber-to-the-home architectures

Fiber-to-the-home network architectures can be divided into two main categories [1]: home run architectures (where a dedicated fiber connects each home to the CO[4]) and star architectures (where many homes share one feeder[5] fiber through a remote node, that performs switching, multiplexing or splitting–combining functions and is located between the homes served and the CO). Star architectures can be active or passive, depending on whether the remote node is

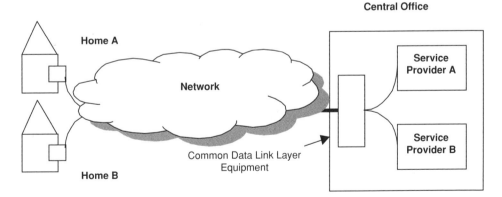

Figure 8.3   Open access-based competition

---

[4] The central office, or CO, is variously called the 'meet point' or 'main node' in contemporary FTTH literature. We, however, will use 'CO' in this chapter.

[5] The feeder loop is the portion of the local loop between the CO and the remote node. The distribution loop is from the remote node to the terminal, while the drop loop is from the terminal to the home.

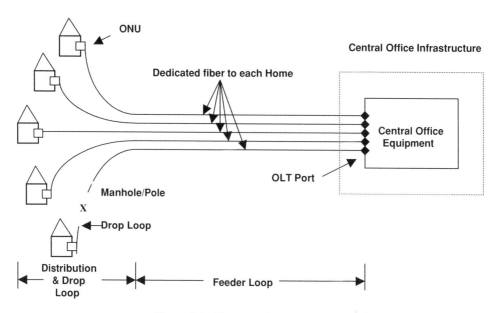

Figure 8.4   Home run fiber architecture

powered or not. Further, the passive star can be a single wavelength system (all homes served by a common wavelength[6]) or a wavelength division multiplexed (WDM) system (where each is served by a different wavelength). Below we examine home run fiber, the active star, the passive star (more commonly known as the passive optical network, or PON) and the WDM PON.

Each feeder fiber is terminated at the central office (CO) on equipment known as an optical line termination (OLT) unit. The CO OLT equipment can be designed to support various data link layer interface types and densities: 100 FX Fast Ethernet, SONET, ATM, and Gigabit Ethernet, among others. On the service provider side, the CO equipment has multiservice interfaces that connect to the public switched telephone network, IP routers/ATM switches or to core video networks [2].

The customer premises equipment (CPE), also known as the optical network unit (ONU) has POTS (plain old telephone service) and 10/100 Base-T Ethernet interfaces and, optionally, an RF video interface. The upstream data and voice signal generally uses the 1300 nm window (1310 nm), while the downstream signal uses the 1500 nm window (1490 or 1510 nm) [2,3]. Broadcast analog video can be delivered (in PONs or in home run architectures) over a separate wavelength as an analog modulated RF multiplex of channels using the 1550 nm wavelength.

8.3.1 Home run fiber

The home run architecture has a dedicated fiber that is deployed all the way from the CO to each subscriber premises (Figure 8.4). This architecture requires considerably more fiber and OLTs (one port per home) compared to the other, shared, infrastructures [1].

---

[6] It is customary to use two or three wavelengths, even in the so-called 'single wavelength' systems. Later, we describe the use of each of these wavelengths.

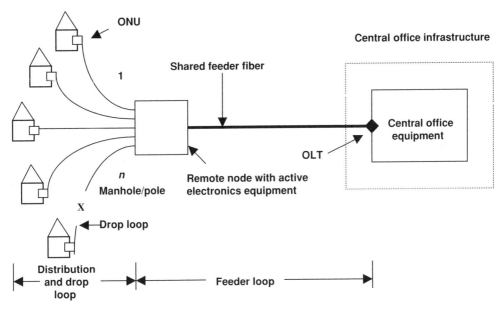

Figure 8.5   Active star architecture

8.3.2 Active star

An active star architecture (also known as a double star) reduces the total fiber deployed, and hence lowers costs through feeder fiber sharing. In the active star, a remote node is deployed between the CO and the subscriber's premises. Each OLT port and the feeder fiber between the CO and the remote node is shared by anywhere from four[7] to a thousand[8] homes (the split ratio) via dedicated distribution links from the remote node [1], see Figure 8.5.

The remote node in the active star network can be either a multiplexer or switch. The remote node switches signals in the electrical domain, and hence OEO conversions are necessary at the remote node [1]. Since the feeder bandwidth is shared among multiple end points, the maximum sustained capacity available to each home – both upstream and downstream – is typically less with an active star architecture than with home run fiber.

8.3.3 Passive star (passive optical network–PON)

In the passive star network, the outside plant has no active electronics (and hence does not need power). At the remote node, a passive splitter replicates the downstream optical signal from the shared feeder fiber onto the (4–64) individual distribution fibers, while a coupler combines optical signals from the individual homes onto the feeder fiber using a multiaccess protocol (Figure 8.6). The OLT allocates time slots to the ONU to transmit upstream traffic [1, 2].

---

[7] In this case, the remote switch is an environmentally hardened device and is mounted on a pole.
[8] In this case, a large cabinet containing the active electronics is deployed.

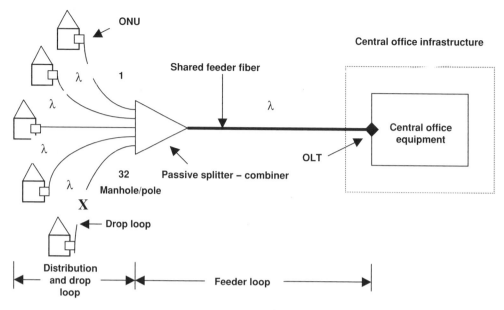

Figure 8.6 Passive optical network

As in home run fiber, in practice, most PON designs use two wavelengths: 1310 nm for upstream traffic and 1490/1510 nm[9] for downstream traffic [2,3]. Generally, the 1550 nm window (1530–1565 nm) is used to provide a WDM overlay for delivering broadcast analog video [2].

8.3.3.1 Design considerations for a PON

A key design consideration for PONs is the location of the splitter. Intuitively, the lowest cost PON architecture is one with isolated pole-mounted splitters placed to minimize the amount of distribution fiber. In this chapter we denote such a PON layout as a 'curbside PON' (Figure 8.7).

Notice that, in a curbside PON, two OLT porsts have to be deployed as soon as the first home in each 32-home 'neighborhood' takes service. Clearly, if we aggregated both splitters at one point (Figure 8.8), and connected fibers to splitters only as needed, we would need to deploy the second OLT only after 32 out of the 64 (or 50 %) of homes subscribed.

The resultant savings from deferring OLT port deployment may be offset by the cost of longer distribution loops (and hence more deployed fiber).

OLT port deployment can also be deferred by distributed splitting. Typically, in a 1:32 distributed split PON, there is a 1:8 (or a 1:4) splitter closer to the CO[10], which reproduces

---

[9] ITU standard G.983 recommends the use of 1310 nm for upstream traffic and 1490 nm for downstream traffic [3]. Some EPON vendors prefer to use 1510 nm downstream however.

[10] In fact it can be located within the CO.

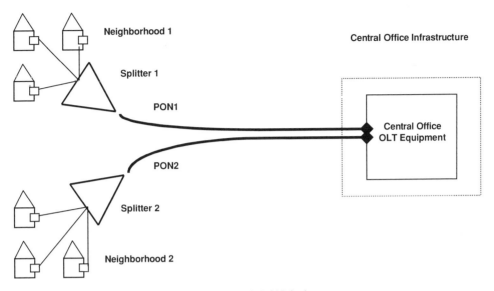

Figure 8.7   Curbside PON deployment

the downstream signal on each of eight (or four) distribution fibers. Each of these eight (or four) distribution fibers, in turn, terminate on a 1:4 (or a 1:8) splitter. Each of these splitters serves four homes (or eight homes), for a total deploy of 8 × 4 or 32 homes. The 'upstream' splitters, if placed in the CO, basically permit homes on different 'downstream' splitters to share the same OLT port (even without any splitter aggregation). These trade-offs (trading off more distribution fiber or distributed splitting or both in order to save on OLT ports) have

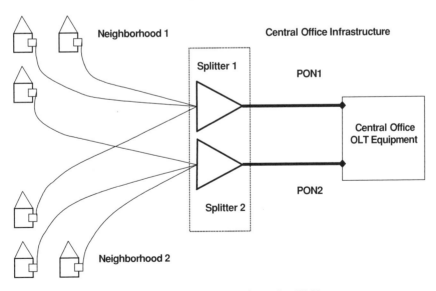

Figure 8.8   Fiber aggregation point (FAP)

been largely unexamined heretofore. We address in Section 8.4.3 how many splitters should be aggregated at an 'optimal' fiber aggregation point (OFAP).

8.3.4 WDM passive optical networks

Wavelength division multiplexing (WDM) is coarse (CWDM) or dense (DWDM), depending on the number of wavelengths multiplexed on to the same fiber. The OLT puts all the wavelengths onto the shared feeder fiber, and the splitters replicate the wavelengths to each home. While CWDM PONs are currently being marketed, DWDM PONs require expensive frequency-stable, temperature-controlled lasers, and are, for the present, not economically feasible, and will not be considered further.[11]

8.4 Economics of fiber-to-the-home

Understanding the cost structure of the industry is a prerequisite to understanding the viability of competition. This section examines the engineering economics of the different FTTH architectures.

8.4.1 Cost model assumptions

It is assumed that the fiber infrastructure is an overbuild in a community already served by copper and coaxial cable. Once the fiber is deployed, video, telephone and data services can be expected gradually to shift to the fiber network. Capital cost per home, as we shall show, is very sensitive to loop lengths (and hence to housing densities), and therefore we consider three deployment scenarios: (i) urban; (ii) suburban; and (iii) rural.

Service providers will find it efficient to lay sufficient fiber during initial construction to support all subscribers in a community, even if only a fraction of them initially sign up for service. The cost of trenching or making poles ready is prohibitively high to go back and retrofit fiber as more homes subscribe to the service. By contrast, the drop loop[12] and the optical networking unit (ONU) can be provisioned as users sign up for service. OLT and remote node electronics can also be deployed based on penetration.

We have chosen three COs in Pennsylvania to represent urban, suburban and rural scenarios,[13] using data from the HAI Model 5.0A (available at www.hainc.com). The HAI Model provides data on the CO area, the number of clusters[14], the radial distance, aspect ratio and location of each cluster with respect to the CO, the total number of homes and housing density for each cluster. Using this data, the costs of deploying feeders and distribution fiber for each scenario can be calculated (Table 8.1).

---

[11] Personal communications with Vendor A reveal that the cost of an 8-wavelength system can be as high as $160 000.

[12] It is not abnormal to preprovision the drop loop as well. In builds where the fiber drop into each home is buried (and especially in new builds) one would in fact expect the drop loop to be provisioned when the rest of the FTTH network is built.

[13] The CLLI codes for the urban, suburban and rural CO are PITBPASQ, HMSTPAHO and TNVLPATA, respectively.

[14] In a cluster, all homes are at a distance less than 18 000 feet from the center of a cluster.

Table 8.1   Deployment scenarios

Deployment	Homes per sq. mile	Homes served per CO	Number of clusters
Urban	5175	16 135	23
Suburban	514	10 183	14
Rural	116	5871	10

8.4.1.1 Capital costs for homes passed

These are capital costs incurred, irrespective of whether homes sign up for service or not, and include: (i) construction costs – the cost of making poles ready (for an aerial build) or the cost of trenching for a buried fiber deployment, and the cost of fiber deployment; (ii) fiber-related capital costs – the cost of feeder and distribution fiber, sheath, splicing and enclosures; (iii) CO related capital costs – the cost of CO real estate, powering, construction and CO fiber termination, (iv) remote terminal related capital costs – the cost of splitter–combiner cabinets for PONs and the cost of remote terminal real estate, powering arrangements and cabinets for active star networks.

8.4.1.2 Capital costs for homes served

Once the fiber is deployed, service provisioning requires deploying networking equipment at the CO (and remote node) and connecting the subscriber to the network by laying the drop loop. The splitters can be prepositioned or incrementally deployed as more homes sign up for service. The central office equipment is organized on racks, with each rack accommodating a fixed number of shelves (usually different for each vendor). Shelves have slots where line cards are plugged in. We assume Ethernet as the data link layer technology.

8.4.2 Cost model results

Figure 8.9 shows the capital cost per home passed for the three architectures for each of the three density scenarios. The capital cost per home served depends quite heavily on the penetration achieved (Figure 8.10).

The necessity of deploying all the fiber up front, with its attendant construction costs, makes FTTH a decreasing cost infrastructure with penetration. The curbside PON appears to be the most economical FTTH architecture. For very low levels of penetration, the home run architecture is significantly more expensive, as more fiber needs to be prepositioned in the home run case, while for high levels of penetration, the cost difference drops to about $200 (at 100 % penetration) per home in an urban deployment.

The cost per home passed (and served) is sensitive to loop lengths, especially for the home run architecture, which has much higher costs than PON in rural areas (especially for low penetration levels).

8.4.3 OFAP as a real option: PON design under uncertainty

For curbside PONs, all OLT ports have to be prepositioned, irrespective of how many homes take service; PON architectures in which splitters are aggregated at FAPs require fewer OLT

Architectures and Deployment Scenarios

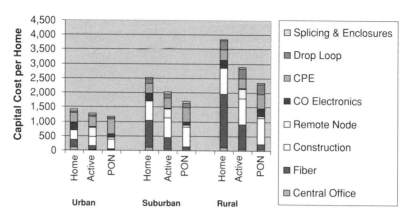

Figure 8.9 Breakdown of capital cost per home for FTTH architectures

ports for penetration levels less than 100 %. In an urban deployment, a fiber aggregation point PON that aggregates four splitters (128 homes) needs 75 % fewer OLT ports compared to a curbside PON at 20 % penetration. Even at a higher penetration (60 %) it requires 25 % fewer OLT ports than a curbside PON.[15]

Capital Cost per Home Served (Urban Deployment)

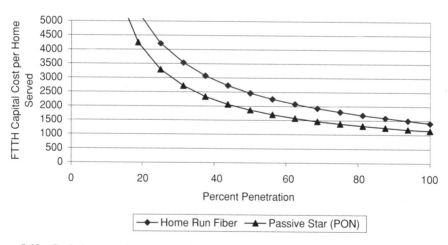

Figure 8.10 Capital cost per home served for an urban deployment (the curve for the WDM PON lies outside the scale chosen for all the plots)

---

[15] For any value of penetration, we assume that the probability $P(n)$, $0 < n <$ (FAP size), that n homes sign up for service is binomially distributed with parameter $p = \%$ penetration. The probability that an OLT port is required for any neighborhood is $1 - P(0)$ in a curbside PON. The approach can be extended for FAP PONs and distributed split PONs.

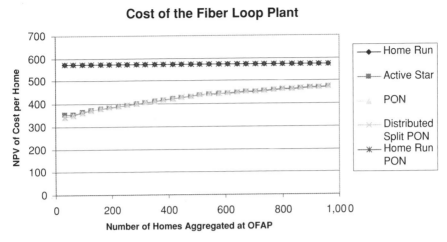

Figure 8.11 Increase in fiber-related capital costs per home as distribution loops are lengthened (urban deployment)

The PON splitter that fills up the last in a particular fiber aggregation point (that has multiple splitters) serves less than 32 homes. Two such splitters (belonging to two different FAPs) that serve less than 16 homes can be served by the same OLT port through a 1:2 splitter placed in the CO. Thus, distributed splitting further reduces the number of OLT ports deployed at low penetration levels. The savings in the central office, however, come at a cost: longer distribution loop lengths. As more splitters are aggregated at a FAP, the distribution loops are lengthened, resulting in higher fiber costs per home (Figure 8.11). Aggregating 960 homes (30 splitters) adds $134 in terms of fiber-related capital cost per home for an urban deployment.

8.4.3.1 Option to defer investment in OLT ports

We now investigate the trade-offs between central office cost savings and increased distribution fiber costs, in order to gain insights into FAP design. Since FAP (and distributed split) architectures sharply reduce the number of OLT ports that need to be prepositioned (vis-à-vis the curbside PON), investment in OLT ports can be deferred until more users sign up. The slower the take-up rate, the longer the investment in OLT ports can be deferred. Using net present value analysis, Figures 8.12–8.13 show that, even for an optimistic take-up rate[16] in the urban context, the lowest cost architecture – taking into account the additional fiber-related costs and OLT port savings – has an OFAP (optimal fiber aggregation point) size of about 200 homes. In rural areas, this is reduced to 96–128 homes. Note that if one further resorts to distributed splitting (in addition to aggregation), the NPV capital costs are even lower; however, there will

---

[16] The optimistic take-up rate scenario assumes that we have 30 % of homes taking service by year five and 70 % of the homes taking service by year ten.

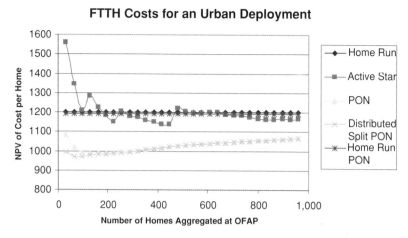

Figure 8.12   NPV of total cost per home for urban deployment for an optimistic take-up rate

be additional operational expense to rearrange splitters at the central office and the OFAP as penetration changes.

8.4.3.2 Option to phase in new technologies and deploy multiple link layer technologies

With technology continuously evolving, one can expect to see next generation PONs with higher OLT port speeds in the future. In a curbside PON deployment, even if one home (among the 32 homes served by each stand-alone splitter) needs to be served by the next generation PON, the OLT port and all the 32 ONUs must be replaced. On the other hand, in an OFAP deployment, while most splitters (and corresponding OLT ports) can continue to

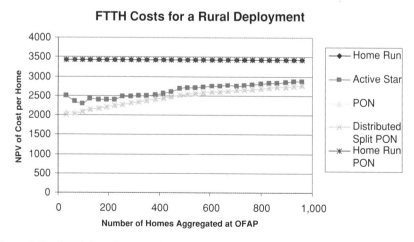

Figure 8.13   NPV of total cost per home for rural deployment for an optimistic take-up rate

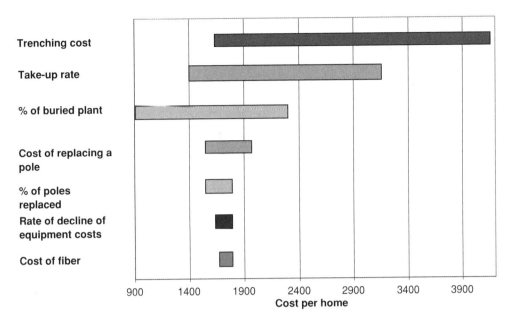

Figure 8.14   Sensitivity analysis for a PON FTTH urban deployment

support the older technology, a few OLT ports can be upgraded to newer technology. Therefore, a service provider that deploys a BPON[17] today can gradually phase in a GPON[18] when that becomes available. Extending this idea further, a service provider could simultaneously deploy different link layers, or group heavy-use subscribers onto a splitter with a lower split ratio. Not only can they deploy an ATM PON and an Ethernet PON simultaneously, but they can also simultaneously deploy PONs that have different OLT port speeds and split ratios[19].

8.4.4 Sensitivity analysis

The discussion on FTTH economics is incomplete without a short discussion on sensitivity. Trenching costs are very uncertain in any FTTH build. The cost of trenching depends, among other things, on the underlying bedrock. Also, in urban areas, restoring the sidewalk and front lawns are additional expenses. Therefore, it is not uncommon to see the costs of buried deployment varying between $25 and $100 per foot. From Figure 8.14, we see that the outside plant cost per home is most sensitive to trenching costs (this is assuming a 50 % buried build). For a PON architecture, a variation in trenching costs can make the cost per home vary by as much as $2504. Varying the take-up rate (at the end of year ten) from 30 % to 90 %, results in a variation of $1776 per home. Since aerial construction costs are cheaper, plants that are 100 % aerial are cheaper than plants that are completely buried, by as much as $1403 a home.

---

[17] A BPON has a downstream bandwidth of 622 Mbps and a 1550 nm wavelength overlay.

[18] A GPON has a downstream bandwidth of 1.2 Gbps or 2.4 Gbps.

[19] For example, a PON that has a downstream bandwidth of 155 Mbps and four splits can co-exist with a PON that has a downstream bandwidth of 622 Mbps and 32 splits.

For aerial builds, the cost of making poles ready and replacing poles (that have no space left on them) can vary between $400 per pole to $5000 per pole, and results in a cost variation of $208 per home. Variation in the cost of decline of optical and electronics networking equipment and the variation in the cost of fiber have only a modest impact.

8.5 Competition, FTTH architecture and industry structure

We now examine competition in the FTTH industry and examine the viability of each model of competition (from Section 8.2) in the context of the FTTH architectures (from Section 8.3) in light of FTTH economics (Section 8.4).

8.5.1 Facilities-based competition

Considering capital costs only, FTTH is a decreasing cost infrastructure (Figure 8.10). Without consideration of operating costs, one cannot say definitively that FTTH is a natural monopoly, but it is clear that large economies of scale and large fixed costs are likely to create significant barriers for a second entrant.

From our cost models, the capital cost per home passed is about $600, and the sunk cost to serve an urban community of 16 000 is $9.6 million. Average revenue per subscriber per month can be assumed to be about $130 (assuming that each home subscribes for voice, video and Internet services) and further assuming that direct costs are about 50 % of revenues, the gross monthly margin is about $65 per subscriber[20]. In order to cover plant costs and the cost of electronics for all subscribers that are being served with its net revenues (in five years[21] with an IRR of 20 %), a provider needs a penetration rate of about 35–40 %. Notably, our simple calculation shows that if a 35–40 % penetration is required for profitability, then in the long run, at most two firms can profitably serve the same market.

In suburban, small town and rural areas, where the supply-side economics are even less attractive, it is difficult to imagine that there will be more than one FTTH firm in the long run. In general, therefore, facilities-based competition in FTTH seems very unlikely.

8.5.2 Competition at the optical layer

Natural monopoly at the physical layer does not preclude competition at higher layers. There are two models for how a single physical facility could support multiple competitors at the optical layer.

8.5.2.1 The wavelength per service provider model

Multiple providers can simultaneously rent different wavelengths on a physical fiber owned by a different party. Each service provider could offer data, voice or broadcast video services with a data link layer technology of the provider's choice on its wavelength. While the WDM PON and

---

[20] This monthly margin goes towards paying back the infrastructure cost as well as meeting operations costs, though here, we assume that the entire amount goes into paying back the infrastructure cost.

[21] is reasonable to assume that private players expect pay-back time horizons of five years or less and IRRs of 20 %.

the home run architecture support this model of competition, single-wavelength systems, like the PON and active star, do not facilitate this model of competition. Implementing competition at the optical layer for a WDM PON would require each feeder fiber to terminate on a port in an arrayed waveguide grating (AWG) that routes each wavelength to the appropriate service provider OLT. Under this arrangement, at most two or three[22] providers can be supported using CWDM, which would not require frequency-stable temperature-controlled lasers, both for the OLT and ONUs. Implementing this on a home run system is unnecessary, as each dedicated fiber can be connected directly to the OLT of the desired service provider.

8.5.2.2 The wavelength per subscriber model

Each subscriber can be served on a different wavelength. Given the number of wavelengths required (equal to the split ratio), this amounts to dense WDM, implying the use of expensive frequency-stable temperature-controlled lasers in both the OLT and the ONU. Needless to say, the active star architecture does not support this model of competition either. To implement this model of competition on a WDM PON, a wavelength router is needed, in place of the AWG, to route each wavelength to the desired service provider. Each home uses a different wavelength, requiring the ONUs to support different wavelengths, creating an inventory management problem in the absence of variable wavelength lasers. Note that in this context, a WDM PON closely resembles the home run architecture, in that each user's traffic is isolated on a unique wavelength. The DWDM overlay in effect creates a 'virtual' dedicated point-to-point facility over the shared PON architecture. However, for the home run architecture, each subscriber's fiber can be directly connected to the OLT of the desired service provider, and the use of multiple wavelengths is unnecessary.

DWDM overlays are economically infeasible today in the access space, and hence we close the discussion on WDM PONs and wavelength-based competition at this point.

8.5.3 Data link layer (UNE-based) competition

If the FTTH physical plant is amenable to unbundling, competitors can rent the fiber as a UNE (unbundled network element) and choose the link layer technology to be used over the physical medium. Providers could use ATM, SONET, Ethernet or Analog modulated RF carriers as their data link layer technology. Since all users served by the same splitter–combiner on a curbside PON (and by the same remote node in an active star architecture) have to be served by the same data link layer technology, a curbside PON physical plant cannot be unbundled, and therefore, this model of competition is not possible in curbside PONs and active star architectures. In the case of the home run architecture, this is easy to implement by directly connecting each subscriber's fiber to the OLT of the desired data link layer service provider.

An OFAP architecture, besides having a lower NPV cost than a curbside PON, also enables data link layer competition. Aggregation of many splitters at an OFAP does not require all homes served by the same splitter to be in the same neighborhood. Splitters at the OFAP can

---

[22] Indications are that not more than 2–3 competitors can be supported using CWDM. PON equipment is economical because it does not use very sophisticated lasers. This requires sufficient isolation between wavelengths and may limit the number of wavelengths to 3–4 on each PON. If each competitor uses two wavelengths – one for upstream traffic and one for downstream traffic – it will be difficult to have more than two competitors.

be assigned to different service providers served by different OLTs at the CO, using data link technology of the competitor's choosing. The number of competitors is limited by the number of splitters at the OFAP – generally 6–8. Aggregating too few splitters not only raises costs but also reduces the potential for data link layer competition.

Clearly, if there are multiple link layer service providers, some of the savings from deferring investment in OLT ports will be lost; in fact if there are as many service providers as the number of splitters aggregated, all OLT ports will need to be prepositioned. This cost of competition depends on the extent to which investment in OLT ports could otherwise be deferred. If the take-up rate is high, the impact on costs will be modest.

8.5.4 Network (and higher) layer based (open access) competition

Different Internet service providers (ISPs), telephone service providers and switched digital video providers can use traditional 'open access' to provide data, voice and switched digital video services. There are two possible models:

1. The service provider can wholesale transport from the data link layer provider and resell service bundles to the subscriber. Each subscriber in this context has only one dedicated ISP. This is typical of current DSL and cable open access arrangements [4].
2. The data link layer provider can sell unbundled transport direct to the subscriber. The subscriber can make separate agreements with one or more service providers and can select an ISP on demand using switching/routing technology provided by the data link provider [4]. This is similar to today's dial-up ISP access model.

The existence of thousands of ISPs today indicates that the barriers to entry for service providers are considerably less than at the facilities or data link layers. Data link layer service providers, can, through vertical integration, leverage economies of scale in facilities or the data link layer to uneconomically limit competition at the services layer.

8.5.5 Why UNE-based competition may be preferable to open access-based competition

Open access based on competitive provisioning of voice, data and video services over a shared transport network is made possible by 'unbundling' the network at the 'logical layer' and the 'resale' of data link layer services. In the absence of dark fiber unbundling, service providers and subscribers are obliged to rely on the data link technology choice of the facilities/data link monopoly, even if they would have preferred an alternate data link technology, such as ATM versus Gigabit Ethernet. In effect, the natural monopoly of the physical layer is extended to the data link layer.

The absence of competition at the data link layer may limit the pace of data link layer innovation. More importantly, voice, video and data service possibilities may be limited by the capabilities of the chosen data link layer. For example, if the network owner selected a PON that does not support a video overlay at 1550 nm, service providers are precluded from offering analog broadcast video services.

Finally, open access competition depends on the data link operator provisioning and policing the quality of service provided to each network layer competitor. With data link competition, QoS of each service provider is essentially independent and under the competitor's control.

Link layer competition is not just about ATM vs. Ethernet. The link layer also defines the port speed (downstream and upstream capacity), the number of splits (in a PON) and therefore, in effect, bandwidth per home. So, one can imagine that with UNE-based competition, different competitors can provision the same flavor of data link technology (whether APON, EPON, BPON or GPON) with different upstream and downstream capacities and split ratios. It is conceivable that competitors may even choose to deploy an active star, while a competitor deploys a PON (for example using hardened electronics at the OFAP cabinet).

In the event that the loop architecture facilitates data link layer competition between multiple players, each data link layer service provider could either choose to integrate vertically with higher layer service providers (like ISPs or video service providers) or choose to provide open access. Vertical integration would permit only as many higher layer service providers as there are data link layer service providers. Whether this is a 'sufficient' number of ISPs or video service providers depends on precisely how many data link service providers can be supported, and on second mile costs, operations and marketing costs, all of which we have yet to consider. If the number of link layer competitors is small, it may be desirable to have open access as well, to ensure maximum competition at the services layer.

8.5.6 Necessary conditions for competition in FTTH

Though a loop architecture that facilitates competition is a prerequisite to competition in FTTH, it is not a sufficient condition. The feasibility of competition in the 'last fiber mile' also depends on second mile costs, ownership and industry structure and community (or market) characteristics.

8.5.6.1 Second mile costs and market characteristics

The costs that we have accounted for in our economic model are only the loop infrastructure costs and data link layer networking costs. However, when services are provisioned over this infrastructure, there are expected to be significant costs related to transporting voice, video and data services to the CO from regional nodes. These costs are known as 'second mile' costs (the FTTH network being the 'first mile'). Second mile costs vary tremendously, depending on the location of the community being served. Evidently, second mile costs are expected to be lower for urban communities and can be sufficiently high for certain rural communities or small towns that competition may not be feasible, regardless of the choice of architecture. An examination of second mile costs is an important next step for this research.

Community characteristics, such as housing density, have implications for cost, as local loop lengths are directly related to housing density. The community size determines the number of homes that a particular CO can serve. Since our cost models indicate scale economies in FTTH deployment, smaller communities would have higher per home passed (or served) costs. Consequently, a smaller community would be likely to support a lower number of service providers. The income distribution of a community, and thus the market demand for services, also has implications for viability of competition in a particular market.

8.5.6.2 Ownership, industry structure and competition

Since FTTH is a decreasing cost infrastructure, the most likely outcome is that there will be only one fiber per home. This fiber can be regarded as a bottleneck infrastructure. Therefore, entry into the services market by a large number of providers is likely to require access to unbundled elements supplied by the owner of the fiber infrastructure and/or open access. Experience from the local telephony industry[23] indicates that a vertically integrated entity that owns the infrastructure and provides services is unlikely to emerge as an efficient, cost-based supplier of network elements to retail competitors.[24] Indeed, experience suggests that perhaps no amount of regulation – with the exception of total structural separation – can provide a level playing field to non-facilities-based competitors.

Beard, Ford and Spiwak further argue in [5] that a vertically integrated entity with a large retail market share will have even more incentives to discriminate against rivals in the wholesale market. When a vertically integrated firm that has a large retail market.[25] share rents out network elements to a retail market competitor, it is very likely that it loses a customer in the process (to the competitor) and the retail margin accruing from the customer. The opportunity cost facing this firm is therefore the average cost of production of the loop and the expected value of the retail margin that may be lost.[26] Therefore, the incentives to supply the 'wholesale market' at cost-based prices, thus facilitating competition in the 'retail' market, are inversely related to the market share of the firm in the retail market.

8.5.7 Industry structure, fiber ownership and competition

Accordingly, given the existence of these discriminatory incentives and the economics of the fiber-to-the-home industry, the most viable long-term competitive market structure involves the presence of a wholesale supplier (that is not vertically integrated) and its efficient functioning as a regulated common carrier.

The presence of a 'neutral' firm that builds and owns the fiber infrastructure and offers nondiscriminatory access to all service providers will significantly lower entry barriers to firms intending to provide video, voice and data services. Since this 'neutral' firm will not provide retail services, it would have no incentives to raise a non-facilities-based service provider's key input of production by non-price behavior. Consequently, the exclusively wholesale and neutral nature of such a firm would permit a market – that could have otherwise sustained only one (or at the most two) facilities-based competitors – to sustain multiple service providers.

We now explore who might build and own FTTH infrastructure and the implications of different ownership scenarios for competition.

---

[23] Charles H. Helein, 'A Call to Arms to Local Competitors', http://www.clec-planet.com/forums/heleinjune14.html.

[24] In this context, a 'retail' competitor is a non-facilities-based competitor providing telecommunications and information services to each home.

[25] In this context, the 'wholesale' market is where the infrastructure owner rents out network elements so that non-facilities-based competitors can provide telecommunications and information services in the retail market.

[26] Opportunity cost = AC + (MS)∗(γ); where AC = average cost of production, MS = market share and γ = retail margin. The opportunity cost goes up with retail market share. Intuitively, this means that the higher the retail market share of the firm, the higher the probability a UNE sale represents a lost retail customer. Conversely, in the presence of infrastructure competition (e.g. from cable), a UNE sale can increase scale economies in infrastructure, thus raising the profitability of the firm's infrastructure, whether leased to retail competitors or used for the firm's own retail operations.

8.5.7.1 Private enterprise

Private players own most of the current FTTH deployments in the United States. Many ILECs, CLECs and Cable MSOs are in the process of making fundamental choices about technology and planning deployments. Private players are expected to build the lowest cost networks and networks that facilitate as little competition as possible. Though one can imagine private players (like electricity or gas companies) playing the role of a neutral infrastructure owner, ILECs, CLECs, Cable MSOs and other overbuilders who own the fiber infrastructure are expected to be vertically integrated service providers as well. This does not augur well for services competition above the facilities layer.

8.5.7.2 Subscriber (or community)

There have been a few suggestions in contemporary literature [6] that, just as subscribers own their home networks, they should own the fiber from the home to the CO. There are in fact new housing builds where builders are contemplating building a fiber to each home, where the fiber is owned by the Homeowner's Association. This green field deployment can lead to a much lower cost, as trenching can be accomplished before roads are paved. Though one can imagine subscriber ownership in green field contexts, it looks very unlikely in current developed residential neighborhoods. One can expect practical problems associated with getting all homes to participate. Even if subscribers were to build and own their fiber, there have to be special arrangements for maintenance of the fiber.

8.5.7.3 Local government

The local government, on the other hand, looks reasonably well-positioned to build FTTH infrastructure [6]. Local governments of many cities have evinced strong interest in building FTTH infrastructure in order to attract hi-tech investment, and many FTTH deployments to date are municipally owned, either directly or through a municipally owned electric utility. Government ownership of FTTH infrastructure can provide the neutral platform over which the private players can provide services. In many communities, the public sector is a large consumer of bandwidth, therefore it seems reasonable for the local government to build this infrastructure. The involvement of the local government can lead to an early and widespread deployment (contrary to the 'cherry picking' that the private players are expected to engage in). Local governments also have easy access to rights of way and, depending on how the project is financed, can also have access to low-cost capital. By limiting its activities to building, owning and maintaining the fiber, and with the private players owning the end electronics, the local government does not have to keep pace with electronics technology that is changing rapidly. Therefore, a public–private strategic partnership seems like one possibility that can lead to a competitive industry structure.

8.5.7.4 Investor-owned regulated common carrier

A final ownership possibility is that of an investor-owned common carrier that is rate-of-return regulated. One particularly interesting case is if private players (who intend to provide service) form a consortium that builds and owns the fiber. The involvement of private players (who intend to provide services) in the shareholding of the firm that owns the infrastructure ensures that the firm has little incentive to vertically integrate and provide services in competition with its owners. If this consortium is regulated, this alternative can also potentially lead to a viable long-term competitive market structure.

8.5.7.5 Migration to desired industry structure

Most telecommunications markets in the United States presently have the following fixed infrastructures: the public switched telephone network (PSTN) and the cable infrastructure, owned by the ILECs and Cable MSOs respectively. The assumption of oligopoly rents (or for some services, monopoly rents) accruing to the network owner have increased the valuation of these network assets considerably, as merger and acquisition activity in this industry has duly reflected time and again. An arrangement that lowers the barriers to entry and promotes competition will have a dramatic impact on these valuations. Therefore, it should come as no surprise that incumbents are likely to oppose any industry structure which reduces barriers to entry by service providers, and to lobby aggressively to frustrate any migration towards such a structure. Indeed, the ILECs have successfully persuaded the FCC that they should be free of any obligation to provide UNEs over newly constructed FTTH networks [7], and the FCC has a pending Notice of Proposed Rule Making which would relieve them of providing open access as well [8].

8.6 Summary

Today, apart from telcos, municipalities, communities and power utilities are at the forefront of FTTH deployment; mostly with the intention of creating a competitively neutral platform that other service providers can use to deliver voice, video and data services. In a market full of vendors that offer different 'flavors' of FTTH technology, most FTTH infrastructure builders (like municipalities and communities) face hard decisions when it comes to selecting a platform (architecture and link layer) and a vendor; the decision being especially hard since they have little or no interest (or expertise) in voice, video and data service provisioning. Our work addresses precisely that predicament: we submit that infrastructure builders should build FTTH infrastructure that is technologically and competitively neutral; where voice, video and data service providers can deploy the technology of their choice to support the services they plan to offer. This chapter suggests that the natural monopoly is in the FTTH infrastructure only; therefore, while facilities-based competition is unlikely, it is feasible to have service level competition. It further shows that OFAP PON architectures not only have the lowest costs, but also are technologically and competitively neutral, in that they support both UNE-based competition and open access. OFAP not only allows higher layer service providers to deploy different types of PON with different link layers and port speeds simultaneously, but also provides an economical real option for the deployment of point to point (active star and home run) facilities as well.

References

1. Reed, D. (1992), *Residential Fiber Optic Networks: An Engineering and Economic Analysis*, Artech House, Boston.
2. Kramer, G. and Pesavanto, G. (2002), Ethernet Passive Optical Networks: building a next generation optical access network, *IEEE Communications Magazine*, **40**(2), 66–73.
3. Klimek, M. (2002), ATM passive optical network, *Alcatel Telecommunications Review*, **4**, 258–261.

4. O'Donnell, S. (2000), Broadband Architectures, ISP Business Plans and Open Access, *28th Annual Telecommunications Policy Research Conference*, Arlington, VA, September 2000.

5. Beard, T.R., Ford, G.S. and Spiwak, L.J. (2001), *Why ADCo? Why Now? An Economic Exploration into the Future of Industry Structure for the 'Last Mile' in Local Telecommunications Markets*, Phoenix Center Policy Paper Number 12, November 2001.

6. St. Arnaud, B. (2001), *Customer Owned, Community & Municipal Fiber Networks – FAQ*, CANARIE, Inc., http://www.canet3.net/library/papers.html.

7. US Federal Communications Commission (2003), *Triennial Review Order*, FCC 03–36, August 21, 2003.

8. US Federal Communications Commission (2002), *NPRM Appropriate Framework for Broadband Access to the Internet over Wireline Facilities*, FCC02–42, February 14, 2002.

Part Three

Technology

9

Backbone Optical Network Design for Community Networks

Ashwin Gumaste[1], Csaba A. Szabó[2] and Imrich Chlamtac[3]

[1]*Fujitsu Laboratories, Dallas, USA*
[2]*Budapest University of Technology and Economics, Budapest, Hungary*
[3]*CreateNet Research Consortium, USA*

9.1 Introduction

The growth of community and metropolitan area networks has increased manifold in the last five years or so and is expected to double every 3.5 years. This segment represents an emerging market for provisioning a range of broadband, enterprise, voice and storage services. If we broadly divide high-speed networking into three classifications of long-haul, metropolitan and access/first mile networks, then the backbone for community networks fits perfectly into the metropolitan area. The metropolitan area is an emerging business sector, mapping multiple services to high-speed network infrastructure. Topologically, backbones for community networks are generically spanned across rings, and the choice of transport is relegated to the optical fiber. The choice of optical fiber is appropriate, on account of the vast bandwidth it provides, in addition to the ability to seamlessly scale the system. The scalability aspect of optical fiber-based backbone networks is brought about by the advent of wavelength division multiplexing (WDM) – whereby multiple frequency differentiated channels are made to occupy the near infinite (> 25 THz) bandwidth offered by the fiber. Community backbone/metropolitan area networks (MANs) have two key functions to perform in the hierarchy of providing end-to-end services. The first function is that of transport, whereby high-speed connections or lightpaths [1] are set up between source-destination pairs. The second function is that of data aggregation. Network elements that support the backbone have the ability to aggregate several lower speed traffic streams into a single high-speed connection – thereby saving cost, as well as optimizing network performance. Topologically,

Broadband Services: Business Models and Technologies for Community Networks. Edited by I. Chlamtac,
A. Gumaste and C. Szabó © 2005 John Wiley & Sons, Ltd. ISBN 0-470-02248-5.

backbone community networks and MANs are built on ring topologies. Rings are chosen due to their proven ability to provide resilience, as well as yielding low network costs (as compared to mesh networks). Future MANs are, however, likely to move away from ring topologies to mesh-like topologies, as witnessed in some parts of North America (Verizon) and Asia.

As far as community networks are concerned, the size of the backbone depends on the specific community it would service, and hence typically, such networks have a ring size of not more than 200 km in circumference in North America (Europe and Asia naturally have smaller sized rings). The backbone fiber ring is further connected to the long-haul network – typically in the case of community networks, this would be an ISP gateway connected to one of the many network elements on the ring. The number of network elements residing on the backbone varies from 6–24, and, typically, each network element supports a combination of WDM (for the backbone) and TDM for the access portion.

In this chapter, we will first describe the design of such a backbone network, and then consider the technologies (TDM and WDM) that build such networks. Further, we will consider the amalgamation of the two technologies at network elements by the recently proposed concept of the multiservice provisioning platform (MSPP), as well as its cousin the multiservice transport platform (MSTP). Since the backbone employs a fiber infrastructure, we will then briefly discuss some of the characteristics of optical network design. The key element of the backbone network is the network element residing at the periphery of such rings. Subsequently, we will study the architecture of the nodes that make up the backbone and distinguish the available multiple classifications. Finally, we will consider emerging technologies, such as Ethernet in MANs, as well as Ethernet with MPLS for providing high-quality services in the metro (community backbone) area.

9.2 Design considerations for community metro networks

Network design and optimization is an art-cum-science that leads to stable, convergent and scalable networks. Community backbones are no exception. A typical community network can be considered as a network of networks spanning the entire community and giving access to end users, customers, enterprises and consumers. The design of such a network involves severe planning based on hierarchical network classification. Typically, a community network can be hierarchically classified into two or three sectors – the first/last mile area, the collector/metro access area and the backbone or the metro core area (Figure 9.1). For our consideration we merge the last two aspects, as the only difference between the two is the network size. The first mile, on the other hand, consists of a set of varied and diverse technologies.

Metro networks (core and access), as mentioned before, consider ring topologies as the choice of network deployment. Rings are well known to provide resilience and also lower equipment cost (on account of the dual degree of connectivity). Metro rings have evolved over the last three decades from legacy TDM-based rings to next generation multiservice-capable WDM rings. This evolution has spread across three generations of development. We shall now briefly consider each such technology generation and the intricacies involved.

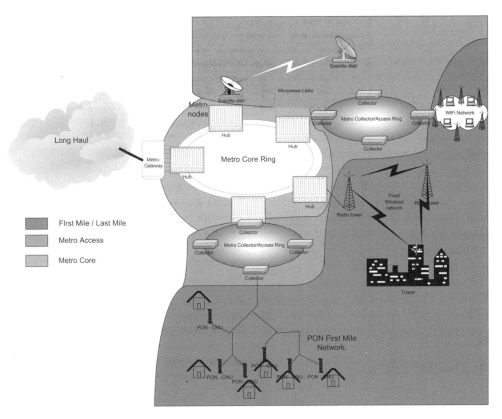

Figure 9.1   Generic community network differentiation and classification

9.2.1 The first generation metro rings

In the early 1980s, Bell Laboratories developed a new transport hierarchy for combining multiple voice streams and enabling the transmission of these cumulative voice streams over fiber optics. This transport technology was based on the well-known principle of time division multiplexing (TDM) of the associated signals in such a way as to achieve a single signal. This led to the establishment of the synchronous optical network, or SONET, hierarchy. SONET networks had the ability to groom slower bit rates of traffic into a high-speed transmission channel using TDM. Further, SONET-based systems had the added advantage of providing monitoring, protection and restoration features through the payload overhead. The development of SONET for the North American market and synchronous digital hierarchy for the European and Asian markets, proved to be a turning point in high-speed information transport. Virtually, or logically, a high-speed SONET line represented multiple lower speed circuits groomed together at network elements. SONET-based systems were broadly deployed in ring network configurations on account of their resilience. In fact, SONET systems were able to achieve 50 ms protection (the ability to heal a fiber cut/node failure in 50 ms). The

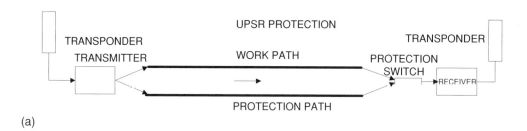

(a)

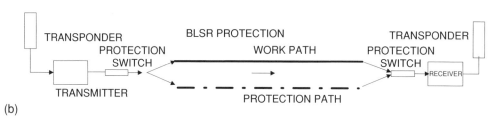

(b)

Figure 9.2 (a) UPSR protection (note that only the receiver has a protection switch); (b) BLSR protection (note that both transmitter and receiver have a protection switch)

50 ms barrier is critical for provisioning uninterrupted voice communication – the primary revenue-bearing SONET service. SONET employs two protection schemes (Figure 9.2(a) and 9.2(b)) called UPSR (unidirectional path switched ring) and BLSR (bidirectional line switched ring).

Figure 9.2 shows examples of UPSR and BLSR. In UPSR system, a protection signal is routed in a graphical complement of the network (as opposed to the shortest path which carries the work signal), and the receiver selects from either the work signal or the protection signal. In contrast, in BLSR systems, the protection signal is dually switched by both transmitter and receiver only when a failure occurs. Naturally, UPSR is faster than BLSR, but it also requires more resources (wavelengths) to function. Figure 9.3(a) and 9.3(b) show the SONET and SDH frame formats. Note that a large portion of the SONET and SDH payload consists of overhead information. This is crucial for signaling and mapping the lower rate flows (called tributaries) into the high-speed synchronous network. Further, SONET and SDH systems are inherently synchronous, meaning that the clock at each node is 'in phase' with every other node. This creates simplicity in data extraction and frame delineation. However, synchronization comes at a price in the equipment. SONET and SDH elements are called ADMs, or add–drop multiplexers. An ADM can drop a particular tributary from a SONET frame, or add a slower rate tributary to the frame. The ADM functions by complete optoelectronic conversion of the signal and further demultiplexing of the complete signal to extract each tributary. The complete reliance on optoelectronic circuitry means that the ADM cost is high. However, this is necessary in order to enable nodes to get access to the lower rate tributary signals embedded in the SONET frame hierarchy. SONET circuits are called optical circuits, or OC-*N*, where *N* is a multiple of the basic STS-1 line rate of 51.84 Mbps. SDH, on the other hand, uses the

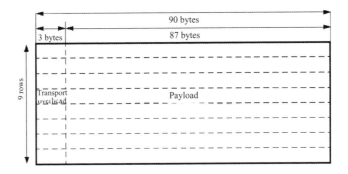

Sonet STS-1 frame format

(a)

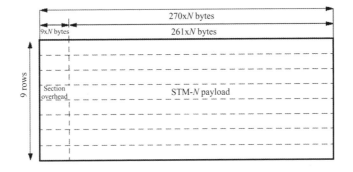

SDH STM-*N* frame format

(b)

Figure 9.3   (a) SONET payload for STS-1; (b) SDH payload for STM-1

convention STM-*N* for its hierarchical deployment, where *N* is an integral multiple of the basic STM-1 speed.

In summary, SONET/SDH networks have been prevalent core technologies for the last two decades or so, and are excellent transport schemes for services such as voice communication. However, the high cost of ADMs to provision SONET streams, and the inability of TDM to efficiently transport bursty IP traffic, means that SONET/SDH hierarchy is unsuitable for next generation Internet and multiservice transport, typically for the community environment.

9.2.2 The WDM solution

SONET/SDH systems can be scaled in the TDM scale only, meaning that, in order to provision a higher speed circuit using SONET/SDH, we have to increase the TDM line rate, i.e. multiplex slower tributaries into the same frame. This seemed to be a plausible alternative for a long while, creating systems with OC-3, OC-12, OC-48 and eventually OC-192 for SONET transmission. However, as the SONET line rate increases, we reach a saturation point whereby electronic systems become more and more difficult to design and are impaired by phase matching and clock skew problems. The limitation in electronics is quite in contrast to the available bandwidth juxtaposed by optical fiber. This gap between the bandwidth provision available by TDM and that available in optical fiber, led to the deployment and advancement of WDM technology. WDM means wavelength division multiplexing, and WDM systems allow multiple SONET or other data streams to co-exist in the same fiber, each residing on a different wavelength. The critical aspect in the design here is to ensure that no two signals have a wavelength (frequency) overlap in the WDM spectra. Please refer to Chapter 4 of [2] to get a more vivid view of WDM systems.

WDM systems currently deploy multiple optical channels or wavelengths, and each wavelength supports an independent communication path. The advent of WDM communication for transport of information was propelled by the advent of two major technologies – arrayed waveguide multiplexers (AWGs), and Erbium doped fiber amplifiers (EDFAs). The former was a subsystem able to multiplex multiple wavelengths (from individual lasers operating at a certain wavelength) into a single fiber stream (as shown in Figure 9.4). Note that the mirror image of the multiplexer was the demultiplexer capable of demultiplexing a composite WDM signal into its constituent wavelengths. The wavelength spacing is predetermined by a standardization movement initiated by the International Telecommunication Union (ITU), and is called the ITU wavelength grid. Current specifications enable 100 GHz (0.8 nm) and 50 GHz (0.4 nm) spacing between adjacent channels, though laboratory trials of 12.5 and 6.25 GHz spacing have also successfully been carried out. The EDFA was developed in the early 1990s in a parallel, independent effort by Myers in the UK and Kinoshita in Japan. The EDFA could simultaneously amplify a group of channels in the optical medium without converting any of the channels into electronic signals. This type of all-optical amplification propelled the rise of WDM for metro and long-haul transport. Some other component technologies that enabled WDM communication are listed in Table 9.1.

9.2.3 The Ethernet solution

Due to its simplicity and cost effectiveness, Ethernet has grown to be a de facto deployment methodology for local area networks. It is estimated that there are some 470 000 Ethernet-based LANs across the globe. One of the primary reasons for the success of Ethernet technology is its adaptability to suit the PC revolution, as well as the ease of mass deployment. With the emergence of the Internet, it was debated whether Ethernet would be adaptable to WAN systems. This led to a serious debate on whether Ethernet technology was disruptive in nature or evolutionary in nature, particularly as compared to legacy SONET systems in the wide area. After much debate, finally, 1 Gigabit Ethernet, and later 10 Gigabit Ethernet, were standardized for WAN deployment. While these standards (under 802.3ae and 802.3z) were conceptually and functionally different from the original Ethernet standard – that deployed carrier sense multiple access (CSMA) for communication, the present GigE (short for Gigabit Ethernet) and 10 GigE (likewise for 10 Gigabit Ethernet) utilize switched Ethernet. This means GigE and 10 GigE are point-to-point connections, and further, they do not require the critical

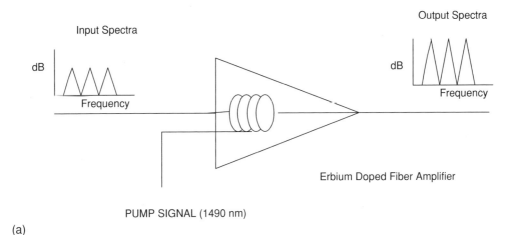

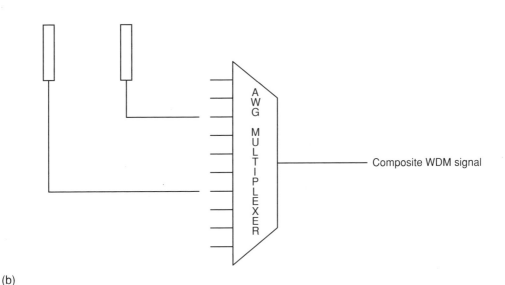

Figure 9.4 Enabling technologies for WDM transmission. (a) Arrayed waveguide multiplexer; (b) EDFA

synchronization that is so necessary in SONET networks. This basic shift from LANs, and cost alleviation as compared to SONET, is one of the key reasons for the success of GigE and 10 GigE.

Performance wise, GigE represents the simplest implementation of a point-to-point connection in an optical domain. This means that a lightpath that is based on GigE protocol carries

Table 9.1 Enabling WDM technologies

Component	Function	Principle	Position in WDM network
Laser	Transmits light on a particular wavelength	Distributed feedback (DFB), or Fabry–Perot cavity lasers are well known	At transmitters and transponders
Modulators	Modulate the emitted light from a laser with the information bearing signal, thereby coupling the information onto the light	Two types are available: external and direct modulators. Use Lithium Niobate substrates for modulation, also available in electro-absorption type	Used in transponders
Photodetectors	Detect the light and generate a current proportional to the information that is carried in the optical line. These need narrowband optics (single wavelength) for functioning	Two types are available: PIN and APD. The PIN photodetector is a simple PN junction separated by an intrinsic region on which photons fall to create electron flow. The APD, or avalanche photodiode, uses the avalanche effect, thereby magnifying the number of electrons that flow through a junction per photon hit on the junction	At transceivers to receive the signal
Optical Switches	Switch optical wavelengths. Can be of add–drop type (add and drop signals or allow to pass through), or can be of optical cross connect type (OXC) cross connect between multiple ports	Mechanical switches: use mechanical fiber movement MEMS (microelectro mechanical switches) use nano structures to move between ports SOA (semiconductor optical amplifier) based switches use an active semiconductor device to switch between fibers. Electro absorption and Mach Zehnder interferometer based switches are also known	Used in optical add–drop multiplexers to drop and add channels or to enable optical cross connect functionality

Continued

Table 9.1 Enabling WDM technologies

Component	Function	Principle	Position in WDM network
Amplifiers	Amplify signals	Erbium-doped or Raman amplifiers known	Used in a distributed fashion throughout the network to enable amplification
GBIC Gigabit interface converters, SFP, XFP	Low-cost standard transmitters and receivers	Distributed feedback lasers or, in some cases, DBR (distributed Bragg reflector) lasers. The receiver is usually of APD type to create a good power budget (due to good APD sensitivity)	These are hot pluggable modules. This means that these modules can be swiftly plugged into network element cards and removed without need for software provisioning
Wavelength blocker	Blocks an arbitrary set of wavelengths from a composite WDM signal, allowing the rest of the wavelengths to pass through	Based on MEMS or liquid crystal type	Used in DOADM applications (dynamic optical add–drop multiplexer)
Wavelength selectable switch (WSS)	Selects a single or a set of wavelengths and switches these to any of the output ports. Comes in $1 \times N$ combination typically, but other combinations are possible	Based on MEMS and AWG. The device dynamically switches any desired combination of wavelengths to any output ports	Full dynamic OADM

GigE frames from a source node to a destination node without any exception. This end-to-end connection often involves cutting through multiple intermediate nodes, all optically. However, this does not mean that GigE has no provision for the aggregation of lower speed streams into itself, like that seen in SONET/SDH networks. At the periphery of networks, Gigabit Ethernet and 10 Gigabit Ethernet switch interfaces often allow slower speed (10/100 Fast Ethernet or other such) signals to be statistically multiplexed to form the mainline GigE/10GigE signal. However, unlike in SONET, where the multiplexing is pure TDM-based, in GigE and 10 GigE, the multiplexing of slower speed signals is based on statistical TDM (STDM). STDM is a low-cost multiplexing technique, but ends up compromising on the Quality of Service (QoS) parameter of the individual ingress streams. The reason for this compromise is the best effort multiplexing scheme achieved through the use of STDM. To provide QoS for the low-cost GigE and 10 GigE transmission versions, a new scheme was proposed – to use MPLS (multiprotocol label switching) along with GigE and 10 GigE, in order to differentiate the traffic flow at switching nodes. However, this also led to cost and management issues. The cost issue is due to the increase in complexity at ingress and egress MPLS switches, as compared to the traditional GigE/10 GigE switches. While the management problems come from having an in-band management system (both data and control in the same plane), leading to issues of severe reliability compromises.

Despite these natural shortcomings, there is still a tangible cost saving in the LAN solution, as compared to the TDM solution. Hence, it is always a difficult choice to decide between conventional TDM and next generation LAN. Of course, both reside on the same optical backbone and use WDM to exploit the near-infinite bandwidth offered by the optical fiber.

9.2.4 Resilient packet rings (RPR)

The IEEE 802.17 standard has been widely accepted as the de facto implementation of the resilient packet rings technology. The RPR system was originated as a packet ring mechanism from two technologies: the IEEE 802.6 DQDB standard and spatial reuse protocol. The RPR system consists of two counterpropagating fiber ringlets, with stations or nodes along the circumference. A station has opto–electro–opto interfaces, such that all the traffic is converted to electronics, processed and then retransmitted. Further, the station has a set of buffers to queue up packets that are yet to be injected into the network. When packets arrive at a station from a local interface, the station momentarily queues up the packets until it finds an empty slot on the ringlet it best thinks will reach the destination. It then transmits the packet on the empty slot. The packet hops from one station to another and when it reaches the destination, it is removed from the network – thereby resulting in spatial reuse of the bandwidth. Protection is carried out by halving the effective bandwidth when a fiber cut occurs. Fairness and QoS can be carried out by implementing fairness control protocols, etc. in the RPR system. For a detailed analysis of RPR refer to [3].

9.3 Functions of the backbone: aggregation and transport – evolution of the MSPP and MSTP concepts

The community network backbone, often spread across the metro access, collector and metro core areas, has two primary functions: aggregation and transport information. While the optical network that comprises the backbone supports the function of transport quite effectively, it is the former function of data aggregation that needs critical attention in core network design. The core represents an opportunistic medium, whereby multiple lower speed connections are grouped together, multiplexed onto a higher speed bandwidth pipe and transported seamlessly from one source node to another. This philosophy entails the optical communication in the core to be relegated to lightpaths residing on wavelengths that have sufficiently large granularity, and further, that ingress nodes that act as sources for these lightpaths groom a large number of slower speed signals onto the lightpath. The criterion for grooming is based on the source–destination proximity requirement. This means that at a source node S, all lower rate streams are groomed onto lightpath L_{SD} say, with the criterion that each stream has destination D in mind. This kind of system is also called single-hop grooming. As opposed to single-hop grooming, we can also have multihop grooming, whereby a stream is groomed into a lightpath that is destined for some intermediate node, and further, the intermediate node has another lightpath destined for the final destination node. In that case, the intermediate node has the function of switching the lower rate stream from the ingress lightpath, to the egress lightpath, as shown in Figure 9.5. Note from Figure 9.5 that the most important element in both single and multihop systems is the digital cross connects. The digital cross connect shown here is a conceptual element with various manifestations in real networks, such as STS-1 to OC–N grooming ($N = 3, 12, 48$ or 192) or 10/100 to GigE grooming. The digital cross connect has

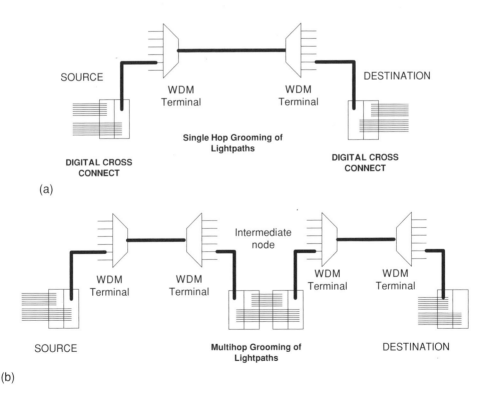

Figure 9.5 (a) Single and (b) multihop systems

the functionality of multiplexing (grooming) slower rate signals into a faster rate signal, and thereby providing the much required high utilization of the available optical wavelength. The principle behind electronic (digital) grooming is to multiplex as many slower rate signals onto one WDM channel as possible, as opposed to sending multiple slower rate signals each on a different WDM channel. This is because the cost of related optics (transponders, etc.) is high, though this economic issue is soon set to reverse with maturity of the technology. However, despite the apparent future reversal, it can be expected that future networks will have a good blend of both digital cross connects and numerous WDM channels.

From the discussion above it is apparent that traffic grooming is an important aspect of network design, especially considering the cost saving achieved through it. However, in real network systems, grooming has one more parameter attached to it, other than just pure traffic aggregation and TDMing – namely service differentiation. Individual data streams each have some service to offer and often function at different data rates. Service differentiation leads to a QoS requirement change in networks. For example, data services are best effort and do not need fast protection, while for voice services, anything more than 50 ms protection is unacceptable. Further, we have a plethora of committed standardized transmission rates for clients to function on, such as OC-3, 12, 48, 192, Fiber channel 1, 2 and 4 G, resilient packet ring speeds, fast Ethernet and GigE/10GigE. Multiplexing all of these on a common platform (if not a common card), and further demultiplexing them seamlessly at destination nodes on the same platform, has been proposed and this functional system is termed the multiservice provisioning platform, or MSPP (Figure 9.6). The MSPP concept provides a platform (hardware implementation) that allows multiple services to be provisioned in a network seamlessly. The last word 'seamlessly' is the key to the success of such a device and that is the essence of the function of provisioning. MSPPs typically have multiple client interfaces that accept lower rate multiple service-based signals. The MSPP then converts these signals into the appropriate network-side (as opposed to client-side) coarse granular signals. The fine granularity of the

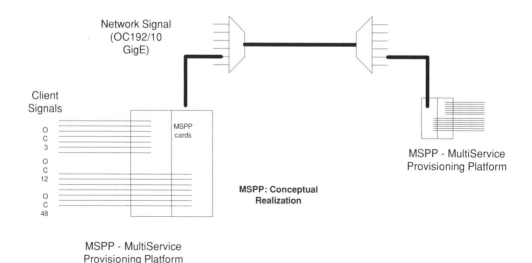

Figure 9.6   Multiservice provisioning platform (MSPP) concept

client-side signal is thus mapped into the coarse granularity of the network-side signal. At the destination node, the MSPP reverses this procedure. It demaps the coarse network-side signal into the finely granular (multiple) client-side signals. In doing so, it may even have to allow some portion of the signal to pass through, aka multihopping. The trick involved in MSPP design is to successfully provision (seamlessly) end-to-end flows from one node to another, such that the client interfaces are completely oblivious of the provisioning aspect. Note here, that client interfaces need to be completely oblivious of the mapping, etc. yet, the MSPP needs to ensure that the service agreement that existed at the time the signal was handed over to the MSPP is maintained throughout the breadth of the provisioning of the network. This aspect of provisioning, along with the ability to enable multiple bit rates and formats to co-exist, has given MSPP its due name and a place of pride in next generation networks, typically for the metropolitan area.

9.4 The optical backbone: design and elements

In this section, we will describe the design and elemental requirements of the optical network that spans the backbone. With the advent of WDM as a technique to maximally exploit the bandwidth offered by the fiber, optical networking has become the de facto standard for network backbones. Community backbones, thus, are no exception. In the community area network, the backbone typically can be mapped into three topologies: single WDM rings, interconnected WDM rings and mesh-based irregular (graph) networks. The choice of network topology depends on three primary factors: the layout of the network elements, fiber rights of way and capacity/service constraints at each network point of presence (POP) site.

Optical ring network design is, at this time, extremely mature, and WDM based add–drop elements have been extensively deployed. These WDM add–drop elements, also called optical add–drop multiplexers (OADMs), have two connections – one into the WDM ring network and one for local access. The local access connection is connected typically to an MSPP platform, as described in the previous section. The combination of the WDM OADM (transport element) and the MSPP is also commonly called the multiservice transport platform (MSTP).

There are several kinds of OADM element to choose from, and this choice generally dictates the functionality of the whole network. Subsequently in this section, we will see the multiple types of WDM OADM presently available in terms of technological demarcation. In contrast to single ring-based WDM networks, multiple interconnected rings face a more significant design challenge in terms of routing and wavelength assignment, as well as interconnection management. For a typical community network, an interconnected ring backbone would spread across a maximum of three linearly connected rings, but typically, most interconnected ring networks have two rings connected. Further, each WDM ring has between 6–24 network elements at its periphery. The size of the ring and the number of elements depends on the optical transmission characteristics of the network, such as span length, dispersion slope of the fiber, node pass-through loss, etc. A way to increase the number of nodes or network size is to induce complete regeneration of the optical signal. This can be done presently by optoelectronic regeneration of the signal, but in the future, all-optical techniques would be possible (2010 and beyond). This kind of complete regeneration is also termed 3R regeneration, in that the signal is reshaped, retimed and reamplified in the process of regeneration. For interconnected rings, the point of interconnection between two rings has two critical properties associated with it:

the interconnection point (hub) can be a single point of failure, and often this can be avoided by redundant hubs (second interconnection point). Secondly, the ring interconnection hub is also a management bottleneck. The following discussion will exemplify the management aspect.

9.4.1 Routing and wavelength assignment: a problem for interconnected ring networks

The number of wavelengths in a WDM system is finite, and hence considered a scarce resource. Further, to establish a lightpath (all-optical connection) on a particular wavelength from a source node to a destination, the primary requirement at the ingress and egress network elements is that the wavelength concerned must be available throughout (across multiple ring spans) this path. This requirement is called the wavelength continuity constraint. This leads to a complex problem, that of assigning wavelengths in the ring for a given traffic matrix. The problem intensifies if we consider the traffic pattern to be temporal (dynamic). The problem becomes even more complicated if we take into consideration the available set of fixed wavelength lasers (within the set of transponders) at a given node. A lot of literature (e.g. [4]) is devoted to the wavelength assignment problem. For interconnected ring networks, the problem is almost unsolvable in real-time (NP-complete). If we consider a path from an ingress node in one ring to an egress node in another ring, then we have to assume that the wavelength should be freely available across the whole path, and further, that the available wavelength has matching transponders (lasers) at the ingress node. This complication is a premier design problem in interconnected ring networks. A real-time solution for dynamic traffic is almost impossible to find, but several heuristics that yield some sort of approximation are available [5].

9.4.2 Mesh optical networks

A mesh can be differentiated from a ring as a network graph where there is at least one pair of nodes that has a degree of connectivity greater than two. Note here that the degree of connectivity means the number of ingress and egress fibers into a node, and for a ring-based network element, this is two. Mesh networks are generally difficult to manage, and current optical mesh level protection is not mature. In fact, most optical mesh backbones use a protection mechanism that is primarily borrowed from the ring topology (shared path or line switched). Typically, backbone mesh graphs have irregular shapes. The motivation for network planners to build mesh backbones is based on the fact that a mesh represents direct source–destination connectivity. At the optical layer, this means more capacity and easier management for routing. However, the protection aspect is in stark contrast. Further, as compared to ring networks, mesh networks have a better wavelength reuse ratio (ability to use the same wavelength in graphically disjoint segments). Architecturally, the primary mesh element is an optical cross connect, or OXC. The OXC has the ability to switch wavelengths from one ingress fiber to another. It may, in the process, also incur some degree of wavelength conversion through optoelectronic means. However, to save costs, all-optical OXCs are preferred. The OXC switching fabric is generally based on MEMS technology. Mesh node design has not yet matured for metropolitan and community backbones. We expect maturity of this technology in the next 2–3 years and deployment beyond that.

9.4.3 Ring network element design considerations

It is estimated that 98 % of community networks currently deploy ring optical backbones, and this high percentage will remain valid for at least another ten years unless there is sufficient technological innovation to provoke change. Ring network elements are a primary design concern for network planners because of the sheer diversity in the alternatives available. It then becomes important to choose the right architectural element, taking into account the plausible alternatives and the requirements the network has, as well as cost. All optical ring elements are variants of the basic optical add–drop multiplexer. The OADM has been a standard system with its own management complex. Ring network management is done through an optical supervisory channel, or OSC. The OSC is an out-of-band control signal that runs through the complete ring. It is demultiplexed at each node through the ring and processed using some modification of an element management system (using, say, a network processor). The OSC is out-of-band with the data. This means that the data wavelengths are spectrally different from the OSC. This leads to reliability – one of the key features desired in optical networks. As a rule of the thumb, it is considered that in-band systems are more scalable than out-of band ones, but when it comes to reliability, out-of-band systems perform far better. Based on this initial description of ring elements, we will now consider three manifestations of ring OADMs: fixed add–drop multiplexers (FOADMs), reconfigurable optical add–drop multiplexers (ROADMs) and, finally, dynamic optical add–drop multiplexers (DOADMs) [6].

9.4.4 Fixed OADMs (FOADMs)

Figure 9.7 shows the conceptual realization of the FOADM. The basic principle of this network configuration is to allow the drop and add of a known, fixed set of wavelengths at a node site. In this case, a dropped wavelength is assigned a particular drop port and this mapping is always preserved.

9.4.5 Reconfigurable OADMs (ROADMs)

Figure 9.8 shows the conceptual realization of the ROADM. This is a popular metro core network element. In the ROADM implementation, unlike in FOADM, any channel from the complete band of WDM channels can be dropped or added. This means that every channel that flows through the system can either be dropped or allowed to continue (or both in case of drop and continue). Further, if a channel is dropped then, due to wavelength reuse, the same wavelength can be used to add another signal into the network at the add side. However, in ROADM systems, each port is mapped to a particular channel. This means that if we have a transmitter at, say, port # 23, and the wavelength emitted by this transmitter corresponds to port # 22, then the transmitter cannot be used in this system. The mapping of wavelength to ports, thus, is always preserved and static.

9.4.6 Dynamic OADMs (DOADMs)

Figure 9.9 shows the conceptual realization of a DOADM. The DOADM is the ultimate in terms of flexibility, as it allows any channel (or group of channels) to be dropped, and further,

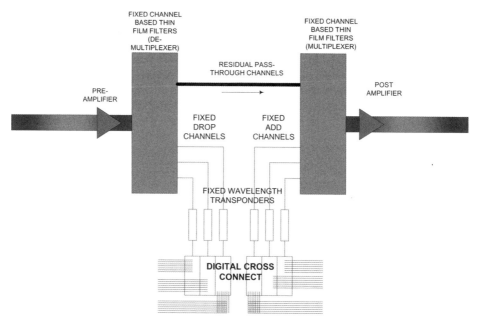

Figure 9.7   Fixed optical add–drop multiplexer (FOADM)

any dropped channel can be dropped to any port. This breaks the mapping between the ports and the channels. Note that for the adding case, the same mapping is no longer needed when using the DOADM system.

9.4.7 Architectural notes on ring OADMs

Figures 9.7 through 9.9 show the three conceptual realizations of the OADMs. Note that the primary difference between the three implementations is in the demultiplexer/multiplexer technology. In the FOADM, the demultiplexing of the signal is done by thin film-based filters.

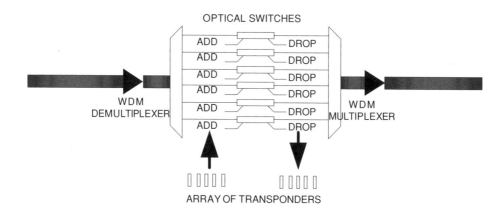

Figure 9.8   Reconfigurable optical add–drop multiplexer (ROADM)

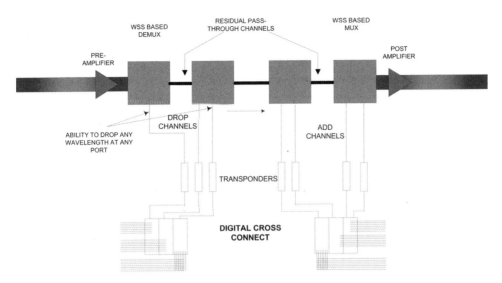

Figure 9.9   Dynamic optical add–drop multiplexer (DOADM)

These are low-cost devices, but are static in nature and also difficult to scale for a large number of channels. Typically, these filters are three-port devices with an input port leading to a filter output port and a rejection (remaining channels) port. In contrast, for the ROADM case, the multiplexer is based on arrayed waveguide technology. AWG technology has been around for a while. This scheme basically multiplexes a group of adjacent (or otherwise) channels into a single fiber. More information on the AWG principle can be obtained in [2]. Finally, in the case of DOADM, we see that the demultiplexer is based on a new switching element called a wavelength selectable switch, or WSS. The WSS is a $1 \times N$ device, whereby an input composite WDM signal is broken into the N output ports in the completely flexible way that the user desires. For example, in one embodiment, the WSS may be switched to drop all even channels at port 2 and odd channels at port 3, in addition to individual channels at port 5 through 8. This kind of complete flexibility allows the network element to break the mapping between the port and the channel.

9.5 Design considerations for the community backbone

The preceding section has discussed the enabling technologies in the core design of community networks. This section shall consider the design of such networks and the considerations that need to be followed for effective design. We will begin by describing the flow chart shown in Figure 9.10. This flow chart gives a high-level insight into the design considerations in a core community network.

The flow chart begins by considering a geographically linked community, and the first task is to predict the user requirements of the community, as projected through the time span of the network being designed. Such quantification can be done by considering the requirement of each user, and further considering the projected requirements to facilitate good futuristic network planning. The user's geographic locations are then mapped.

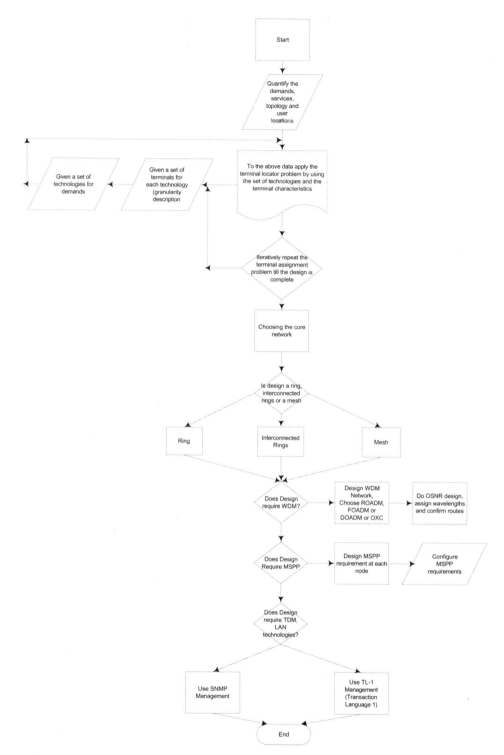

Figure 9.10 Flow chart detailing design considerations of the community backbone

The backbone design then uses the inputs of all the users and applies the terminal assignment problem (a linear program giving the optimum set of backbone POP sites, subject to the traffic and geographic profile). The terminal assignment problem can be found in [7]. By considering the various technologies for the backbone as part of the terminal assignment problem, the LP results in quantifying the requirement of each line card. Subsequently, from the POPs, we can choose the network topology of a ring/mesh or a set of interconnected rings.

Depending on the network capacity, we choose whether to use WDM or not. If we do use WDM, then we design the network based on WDM transmission issues, taking optical signal-to-noise ratio (OSNR) and dispersion into consideration. We optimally place amplifiers, regenerators and dispersion compensators to allow successful communication. This completes the basic optical backbone and we move to the aggregation aspect at each POP site.

The next design step is to decide what kind of aggregation platform is suitable at the periphery of each node. Decisions such as whether to place an MSPP or MSTP are to be made here. MSPP requirements and card count are outlined.

Finally, the services are taken into account. Decisions such as whether to use LAN or TDM technologies form the crux of this design phase. Newer technologies like resilient packet rings (IEEE 802.17– RPR standard) and MPLS are also considered. Based on the services, a management system is chosen and the network design is then simulated for further tests.

9.6 Summary

In this chapter we have discussed community network backbone design. The backbone for community metropolitan networks is based on an optical framework. This framework has multiple manifestations. We have discussed the technologies and different manifestations of the optical backbone. We considered both pure optical and legacy optoelectronic technologies. Amongst the optical technologies we considered are WDM design for rings and meshes, as well as the design of WDM network elements. For optoelectronic technologies, we considered conventional SONET/SDH, the emergence of Ethernet and the advent of the new concept of the multiservice provisioning platform, or MSPP. We also outlined the design principles for future community networks.

References

1. Chlamtac, I., Ganz, A. and Karmi, G. (1992), Lightpath Communication: A Novel Way to Increase the Bandwidth of Optical WANs, *IEEE Transaction on Communication*, **40**(7), 1152.
2. Gumaste, A. and Antony, T. (2002), *DWDM Network Designs and Engineering Solutions*, Pearson Education.
3. Davik, F., Yilmaz, M., Gjessign, S. and Uzun, N. (2004), IEEE 802.17 Resilient Packet Ring Tutorial, *IEEE Communications Magazine*, March, pp. 112–117.
4. Ramaswami, R. and Sivarajan, K. (1995), Routing and Wavelength Assignment, *IEEE/ACM Transaction on Networking*.
5. Ramaswami, R. and Sivarajan, K. (2001), *Optical Networks: A Practical Perspective*, 2nd edition, Morgan Kauffman.
6. Gumaste, A. and. Chlamtac, I. (2004), Light-trails: An Optical Solution for IP Transport, *OSA Journal on Optical Networking*, **3**(5), 261–281.
7. Robertazzi, T. (1992), *Planning Telecommunication Networks*, Prentice Hall.

10

A Comparison of the Current State of DSL Technologies

Scott A. Valcourt
University of New Hampshire, InterOperability Laboratory, Durham, USA

10.1 Introduction

During the telecommunications explosion of the mid 1990s, the vision for an asynchronous last mile link to deliver video services over existing voice-grade cable drove the early investment in Asymmetric Digital Subscriber Line (ADSL). With traffic patterns that mirror those of high-speed Internet access, ADSL and its implementation began with a completely different market than originally envisioned. In this same period, other DSLs emerged, using the same or similar physical line codes with different bandwidth and implementation schemes. Single-pair, High-bit rate Digital Subscriber Line (SHDSL) and Very high-bit rate Digital Subscriber Line (VDSL) are two other DSLs that play significant roles in delivering streams in the last mile. In the next generation usage of the DSLs, the convergence of voice, video and data is a driving factor.

In this chapter, we highlight the major features and benefits of ADSL, SHDSL and VDSL technologies. We address historical and current technological implementations, as well as emerging standardization efforts affecting the DSL technologies, indicating general results of technology testing at the University of New Hampshire InterOperability Laboratory. Finally, we contrast these same DSL technologies as related to key worldwide implementations, and make recommendations on the benefits and cautions when implementing DSLs in the last mile community network.

Telecommunications adoption follows Moore's Law, in that voice services took about 50 years to be accepted; television needed just 15 years; the Internet and related services change the way we communicate daily and take a mere amount of months for widespread adoption. These trends are fueled by a series of perspectives that impact them: substantial

Broadband Services: Business Models and Technologies for Community Networks. Edited by I. Chlamtac,
A. Gumaste and C. Szabó © 2005 John Wiley & Sons, Ltd. ISBN 0-470-02248-5.

growth in the telecommunications market, increasing network complexity, deregulation and privatization, communication convergence and customer orientation. New technological developments and the desire to acquire them led to the expansion of the telecommunications market. In fact, it could be argued that the development and expansion of the private data communication industry spilled over to the telecommunications industry. That spillover caused a demand by users who had become used to fast access to data.

In order to get that access to the widest number of users, the public network needed to upgrade components that are very old with respect to networking in general. Along with these technological advances, much more work in the standardization arena has added to the network complexity. Telecommunications providers are reluctant to add anything to the network without clear assurances that everything in the network will remain backward compatible. As deregulation and privatization continue to emerge since the AT&T divestiture and the Telecommunications Act of 1996, new rules and regulations and new technology offerings add to the general complexity.

With the advent of new services and the desire to implement expensive additions to the public network, telecommunications companies are trying very hard to encourage customers to purchase the service. This influx of new customers is a very strong element; in some cases, the customer group can be just as strong in the policy direction as the Federal Communications Commission (FCC). With local communities considering competitive offerings to move broadband networking services into the mainstream faster than large service providers can deliver, the last mile access is becoming a major element in the development of municipal services.

To address the need for high data rates in the last mile, broadband access technologies like Digital Subscriber Line (DSL) were introduced. What DSL was designed to do, does do, and is being considered to do has an impact on the delivery of broadband services in the community communications setting.

10.2 ADSL vision and history

With traffic patterns that mirror those of high-speed Internet access, ADSL and its implementation began with a completely different market than originally envisioned. Video over ADSL was the first application considered a beneficial match for ADSL. The pattern of requesting a video recording from a central site and its transfer to the subscriber is an asymmetric data transfer. The system requires a small bandwidth level for image requests and a larger bandwidth actual image transfer. This transmission pattern was documented and patented by Bell Atlantic Network Services in 1998 [1].

While such a beneficial means of combining services onto the proverbial single cable was devised, the result was that the marketplace was not yet interested in video delivery via phone line. Rather, in the mid 1990s, organizations like the ATM Forum were interested in the last mile delivery of high-speed Internet data traffic. The World Wide Web was becoming an active force on the Internet, and the traffic patterns of web clicks and graphical downloads matched those of video-on-demand. Therefore, ADSL was developed with a vision of delivering high-speed Internet access, much faster than dial-up modems and not requiring an additional wire at the customer's premises. The Telecommunications Deregulation Act of 1996 aided in the deployment of ADSL in the public switched telephone network (PSTN), because it allowed service providers that were not Regional Bell Operating Companies (RBOC) to have access to central offices and lines on the poles to deliver ADSL services.

Today, ADSL maintains a reasonable stronghold in North America and maintains a commanding market share in Europe and Asia. With regulatory agencies, such as the Federal Communications Commission (FCC), and RBOCs providing more guidelines and details on the service delivery of ADSL, the options available to the end user subscriber become more diverse and robust in terms of bandwidth speeds and value-added service offerings.

10.3 ADSL technology

Originally envisioned as a transport mechanism for video-on-demand applications, and currently positioned as the favored mass-market DSL solution for Internet access, ADSL is capable of downstream rates (towards the customer) from 384 kbps to as high as 8 Mbps in some cases. Downstream reach can extend up to 18 000 feet at approximately 1 Mbps, depending on line quality. In order to achieve these downstream rates, ADSL allocates less bandwidth to the upstream. Typical upstream speeds (away from the customer) are in the range of 64 kbps to 768 kbps, depending on the reach.

ADSL technology, as defined in the ANSI T1.413–1998, ITU-T G.992.1 (G.dmt) and ITU-T G.992.2 (G.lite) standards, uses discrete multitone (DMT) as the line coding protocol. This method, pioneered at Bell Laboratories, divides the frequency spectrum of the line into equally spaced subchannels, often considered to be subcarriers, in that different carrier waves are used to base the coding for each subarea of the frequency spectrum [2–5].

The useful spectrum extends from 0 Hz to approximately 1.1 MHz, being divided evenly into 4.3125 kHz bands, allowing 256 subchannels to deal with interference and rate adaptation. Should there be a frequency interferer in a particular band, the two network ends can negotiate, recognize the interference and agree to disable communication within a specific bin. While maximum theoretical bandwidth is diminished in these situations, even in very noisy environments, ADSL will still function at a reasonably expected level.

Telephone wires are bundled together in multipair binders containing 25 or more twisted wire pairs. As a result, electrical signals from one pair can electromagnetically couple onto adjacent pairs in the binder. This problem is known as crosstalk and can impede ADSL data rate performance. As a result, changes in the crosstalk levels in the binder can cause an ADSL system to drop the connection. Other issues that can cause dropped or impeded connections include AM radio signal disturbers, temperature changes and water in the binder [6]. Additionally, the maximum bandwidth levels that can be achieved in the field are dependent on the cable distance between devices. The longer the cable, the lower the bandwidth levels, due to the increase in signal-to-noise ratio (SNR) resulting from the physical resistance of the medium to the current flow [7].

The key elements of an ADSL network include the central office component, Digital subscriber Line Access Multiplexer, or DSLAM, the cable channel and the customer premises equipment (CPE), or modem. DSLAMs typically multiplex several ADSL modem lines into one channel on the digital side, while splitter technology, either internal or external to the DSLAM, separates the voice channel to the plain old telephone system (POTS), see Figure 10.1. Digital sides typically are asynchronous transfer mode (ATM) devices or Fast Ethernet devices. Modern DSLAM implementations are turning towards the installation and support of Internet

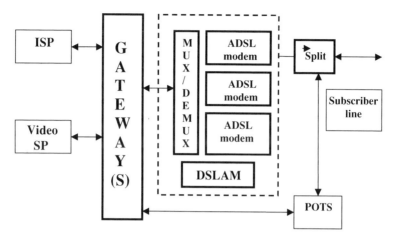

Figure 10.1   Key elements of an ADSL system on the CO side

Protocol (IP) routing capacity within the DSLAM, rather than an additional external device. IP-DSLAMs, as they are commonly called, focus on the delivery of high-speed Internet services by service providers.

CPE modems generally offer connectivity on one side to the DSLAM and the central office, while providing some form of Ethernet interface for the user. Small firewall routers and switches, formerly external and downstream from the modem, are now common within the modem, offering safety and security to the always-on nature of the ADSL service.

Information and data from the DSLAM to the CPE modem is carried via the ADSL super-frame structure. The superframe is a large envelope for packaging 68 ADSL frames and one synchronization frame at a time, all in a period of 17 milliseconds. Forward error correction (FEC) is used to ensure that the data bits transmitted are valid. To further transmit reliable data, two methods of frame transmission are possible: fast data and interleaved data. Interleaved data weaves the bits to be transmitted into the superframe to ensure the likelihood of a large cluster of bits in one frame is not corrupted and lost. Fast data is simply noninterleaved data. Interleaving, while succeeding in reducing potential noise interference at the framing level, increases latency in the transmission of the data. Fast data transmission has very low latency, but has the potential for loss of data due to interference.

Enhancements to ADSL are presently emerging within the technical community. ADSL2 (ITU-T G.992.3 and G.992.4) focuses on enhancing the present ADSL standards by increasing performance in data rates and reach performance over ADSL, and providing for rate adaptation and power management, among other enhancements. ADSL2plus (ITU-T G.992.5) defines the doubling of the bandwidth used for downstream data transmission, achieving data rates in the downstream of up to 20 Mbps on lines as long as 5000 feet. These enhancements are designed with backward compatibility, allowing rollout at one side or the other without requiring a coordinated dual-side rollout [8].

ADSL2 rate and reach performance enhancements approach achievable downstream rates near 12 Mbps and upstream data rates of about 1 Mbps, depending on loop lengths and other factors. This is accomplished by improving modulation efficiency by mandating a four-dimensional, 16-state trellis code and 1-bit quadrature amplitude modulation (QAM) constellations, which provide higher data rates on long lines where the signal-to-noise ratio (SNR) is low. In addition, using frames with programmable numbers of overhead bits can reduce the framing overhead that ADSL standards contain. The ADSL overhead bits per frame are fixed and consume 32 kbps, or one frame of the superframe, of actual payload data. The ADSL2 overhead bits per frame can be programmed from 4 to 32 kbps, providing an additional 28 kbps for payload data.

ADSL2 defines seamless rate adaptation (SRA) as a means of changing the data rate of an ADSL2 connection while in service, without any bit errors or connection interruptions. ADSL2 monitors for changes in the channel conditions and adapts the data rate to the new channel conditions without user intervention. SRA is achieved as a result of the decoupling of the modulation and framing layers in ADSL2, enabling the modulation layer to change the transmission data rate parameters without modifying parameters in the framing layer.

ADSL transceivers operate in full-power mode day and night, even when not in use. With several million ADSL modems deployed throughout the world operating in this 'always-on' model, wasted power consumption was, and is, an operations issue that the worldwide standards organizations needed to address. To address these power issues, the ADSL2 standard provides for two power management modes that help reduce overall power consumption. These modes include the L2 low-power mode and the L3 low-power mode.

The L2 low-power mode takes place at the DSLAM side of the network, enabling statistical power savings by switching to and from low-power mode, based on Internet traffic levels over the ADSL side of the DSLAM. The L3 low-power mode takes place at the CPE modem and DSLAM sides and allows the system to enter into a sleep mode when the communication link is not being used for an extended period of time.

ADSL2plus, as defined in ITU-T G.992.5, doubles the downstream bandwidth, which results in an effective increasing of the downstream data rate on lines less than 5000 feet. While ADSL and ADSL2 define a downstream frequency band up to 1.1 MHz and 552 kHz respectively, ADSL2plus specifies a downstream frequency up to 2.2 MHz. The ADSL2plus upstream data rate is about 1 Mbps, depending on loop conditions.

ADSL2plus provides the capability to use only tones between 1.1 MHz and 2.2 MHz by masking the downstream frequencies below 1.1 MHz. This feature can help binders that are delivering both ADSL and ADSL2plus services minimize crosstalk at both the central office and customer premises ends.

ADSL2 and ADSL2plus provide the ability to channelize bins in the frequency bandwidth areas to support different channels with specific link characteristics. These reserved channels can support applications like videoconferencing, which requires special dedicated channels to ensure all voice and video packets are received. In addition, channelized voice over DSL (CVoDSL) is a means for ADSL2 customer premises to receive more than one voice channel without the need to run more copper pairs to the customer.

At the University of New Hampshire Inter Operability Laboratory, testing of the ADSL technology has been ongoing since 1997, with testing focusing on the ANSI and ITU-T Recommendations. Within the past year, particular emphasis has been on the ADSL2 and ADSL2plus technological advances. Like all technology advances, there exist categories of developers at various stages, loosely falling into early developers, active second tier developers and reseller/original equipment manufacturer (OEM) agreements. In this period of testing, in large group and day-to-day testing, early developers and active second tier developers have participated. Many of the new features described above have been implemented, and several of those features are interoperable within the collective community. With some further technology work still to be done, ADSL2 and ADSL2plus are poised for a rapid deployment of new features and services for the ADSL user community.

ADSL is configuring and delivering the present and future broadband communications for the home user via the one-hundred-year-old POTS cable plant, all without the need to deliver more copper and without impacting the existing voice services in which RBOCs have invested so heavily.

10.4 SHDSL technology

While ADSL provided a beneficial mechanism for broadband delivery in the asymmetric space, the technical community sought a means for replacing existing expensive T1 technology with a symmetric methodology. In North America, the ANSI T1E1.4 Committee began work on a new standard version: High bit-rate Digital Subscriber Line second generation (HDSL2). The HDSL2 standard, captured in the document ANSI T1.418-2000, uses a different transmission method in order to achieve 1.544 Mbps symmetric transmission over the same frequency spectrum. It is possible that it could support rate adaptive services in the future, allowing lower speed access on even longer loops. The new method incorporates pulse amplitude modulation (PAM) line code, a coding scheme, higher transmission power and special pulse shaping for spectral compatibility with other existing services. This method achieves an improvement of up to 7 dB over 2B1Q HDSL, its technical predecessor, and provides cost benefits since only one data pump per modem is required to provide symmetric 1.544 Mbps instead of the original two, thus saving a copper pair in the cable plant. The downside at present is the high complexity and high power required for such a solution [9].

The ANSI T1E1.4 Committee established a clear direction for the evolving HDSL2 technology. The committee agreed to use trellis-coded PAM (TC-PAM) line coding with a modified OPTIS (overlapped PAM transmission with interlocking spectra) spectral shape. The shaping of the HDSL2 signal reduces crosstalk to a controlled level, in order to enable telecom operators to deploy different types of DSL, including HDSL2 and ADSL, on the same bundle of lines. While spectral compatibility is a very important area of study and standardization in the ANSI Committee, HDSL2 eventually became the North American position on the standardized technology called Single-pair High bit-rate Digital Subscriber Line (SHDSL) technology.

ITU-T Study Group 15, Question 4 began work in early 1999 on the ITU-T G.991.2 (G.shdsl) Recommendation, with a final version ratified in February 2001. The work done in ETSI TM6 on the SDSL standard became the European input to the G.shdsl standard, making a symmetric solution that contained a common base worldwide and supported regional differences and technology evolution.

SHDSL is designed to provide a flexible replacement to the worldwide usage of symmetric digital data transmission, by implementing in the standard design several different line rates over the copper communications channel. These line rates coincide with existing line rates in the field, such as T1, fractional-T1, ISDN-BRI, T3, E1, fractional-E1, E3, HDSL, ADSL upstream and downstream, and others [10].

Further, SHDSL does not support a native voice channel in the spectrum, allowing data only transmission to dominate the communication channels between the central office and customer premises equipment. As a result of SHDSL's support of a vast collection of existing and future symmetric line speeds, the technology is dependent on the handshaking standard, ITU-T G.994.1 (G.hs). This handshaking protocol is used in combination with all of the ITU-T G.99x Recommendations to cause the central office and customer premises equipment to determine what devices are on each end of the copper communications channel, and to set the best possible technology and rate attempts that the end devices can support [11].

10.5 VDSL technology

Very high bit-rate Digital Subscriber Line (VDSL) technology is the latest and fastest member of the DSL family to emerge onto the scene. From a core protocol standpoint, there is very little that VDSL has that differs from ADSL. VDSL uses DMT and QAM to support the delivery of frames of data and is capable of downstream rates from 12.96 Mbps to as high as 55.2 Mbps in some cases. Downstream reach can extend up to 4500 feet at approximately 12.96 Mbps, depending on line quality. Typical upstream speeds (away from the customer) are in the range of 1.62 Mbps to 6.48 Mbps, depending on the reach [12].

VDSL operates in the frequency band located between 300 kHz and 30 MHz. Because VDSL has such a short cable distance, and the potential for electromagnetic interference (EMI) in the form of radio frequency (RF) egress noise, tightly controlled limits need to be placed on VDSL network systems in order to prevent them from interfering in the internationally standardized amateur radio bands [13].

However, the perceived installation and rollout of VDSL takes on a slightly modified appearance. Due to the increase in bandwidth, shorter loop lengths are required. With the issues already existing, and being revised, in the ADSL standard, the VDSL technology suggests a new layout, one that supports the use of digital loop carriers (DLC) to remote locations, then the installation and delivery of VDSL devices to neighborhoods. To establish the VDSL remote locations, fiber optic cable would be extended out to remote nodes, with VDSL proposed to make the remaining hop to the subscriber's premises, delivering voice and high-bandwidth data channels over a single twisted pair [14]. The extremely high bandwidths available from VDSL will enable a new generation of broadband applications. Even though the data rate is higher, it is likely that the shorter distance and more controlled transmission environment may make this technology easier (and potentially less expensive) to implement than other technologies.

VDSL technology is being addressed in various standards bodies throughout the world and in various regions. In North America, the VDSL standard is defined in a trial use standard within the ANSI T1E1.4 Committee. ETSI defines the European standard and the ITU is defining the international version of VDSL with much difficulty, as line code choice for VDSL appears to be a matter of regional choice and technical scrutiny. Virtually every combination of line code, specification and access method has been suggested over the last several years, with two major implementations being standardized in various bodies. The camps are so divided that it is challenging to predict which line code will be chosen worldwide, and regionally in some cases. Regional deployments of a pre-standards flavor have already been installed, and much practical research work has already been accomplished worldwide on implementations of the standards.

While VDSL still has a planned track of future development and enhancement, its techno-logical implementation proposes a future that is yet to identify an application that will overtax the technology on the user end.

10.6 The best technology

Table 10.1 provides a comparison of the main technical parameters and typical application areas of the xDSL technologies discussed.

With each DSL technology designed to address a different market segment need, it is chal-lenging to determine the best technology in a general sense. Several marketing analyses and trade journal communications have already made market predictions on the best technology to purchase. However, the best technology must be selected with the user/implementer in mind. Since most home-based or business-based users are dependent on the service provider to implement a technology, choice may be a limiting factor in determining the best technology available. If all technologies were available everywhere at all times (a challenging proposition), then the best technology choice would have as a lowest common denominator that of intended use.

Symmetric networks and asymmetric networks offer benefits that differ. As we have already discussed, traffic patterns and intended application types for a specific DSL technology play a role in the specific technological choice that needs to be made at the time of installation. While individual users may have a clear vision as to the best technology for the intended applications that would be used on this broadband network, service providers do not, and must find means for delivering the greatest services over the broadest needs base [15].

10.7 DSL community networks

DSL technologies, especially in light of the deregulation of the telecommunications sector, provide a significant means of technology delivery to a community user base. Colocation of DSLAMs in existing central offices and access to the actual copper on the poles, versus space on the poles, gives municipalities and communities an opportunity to connect individ-ual homes and businesses on their own, without the implementation overhead that a service provider might introduce. Further, municipalities that implement community networks using DSL may be able to offer a greater cost–service ratio than a broad service provider might introduce.

While the regulatory restrictions have diminished to a lower level, there are regulatory issues that need to be addressed appropriately for DSL technologies to be deployed in a community network. The network deployment must still follow all of the existing telecommunications and regulatory laws and rulings that exist, which means that a municipality may require expert services to deploy a community network of DSL devices. Further, there are power restrictions and costs that such a community network service would need to cover, and if the community base is too small, such costs may not be reasonably covered without major funding.

In the end, DSL technologies offer a good land coverage area, and with DLC and fiber-to-the-neighborhood (FTTN) options emerging in many new construction projects, the ease in reaching subscribers beyond the 18 000 foot limit of ADSL is enticing.

Table 10.1 Comparison of xDSL technologies

Name	Standard	Technology (line coding)	Technology (symmetric/ asymmetric)	Data rates	Typical applications
ADSL	ANSI T1.413-1998 G.992.1, G.992.2	DMT	A	384 kbps–8 Mbps downstream (1 Mbps at 18 000 ft), 64–768 kbps upstream	Web browsing
ADSL2	G.992.3, G.992.4	QAM	A	Up to 12 Mbps (down), Up to 1 Mbps (up)	Video
ADSL2plus	G.992.5	QAM	A	Up to 20 Mbps (down) with up to 5000 ft, 800 kbps upstream	Video
HDSL	ANSI T1.418-2000, G.991.1	PAM	S	784–2320 kbps symmetrical	V/V/D in business applications
SHDSL	G.991.2	PAM	S	192 kbps–4.6 Mbps symmetrical	V/V/D in business applications
VDSL	ANSI T1E1.4/2000	DMT, QAM	S/A	12.96–55.2 Mbps (down) (12.96 Mbps at 4500 ft), 1.62–6.48 Mbps (up), Up to 26 Mbps symmetrical	Broadcast video, VoD

10.8 Summary

While DSL technology continues to emerge with feature enhancements and new protocol and line code applied theories, the number of technologies using the name Digital Subscriber Line also increases. The marketing value associated with such a label is significant on a worldwide scale. Beyond the message that DSL provides, like-minded citizens within a tightly coupled community can adopt a last mile technology, such as DSL, to deliver broadband services specific to the community. With foresight into the regulations and understanding of the technology and the applications expected to operate in this broadband network, community DSL networks are emerging as a viable communications medium for the modern community.

References

1. Seaholtz, J. W., *et al.* (1998), United States Patent Number 5 812 786.
2. Summers, C. K. (1999), *ADSL: Standards, Implementation, and Architecture*, CRC Press.
3. ANSI (1998), *Network and customer installation interfaces – Asymmetrical Digital Subscriber Line (ADSL) Metallic Interface*, Standard TI. 413–1998.
4. ITU-T (1999), *Asymmetric Digital Subscriber Line (ADSL) transceivers*, Recommendation G.992.1.
5. ITU-T (1999), *Splitterless Asymmetric Digital Subscriber Line (ADSL) transceivers*, Recommendation G.992.2.
6. Venugopal, P. (2002), *Radio Frequency Interference and Capacity Reduction in Digital Subscriber Line Technologies*, Master's Thesis, University of New Hampshire.
7. Busby, M. (1998), *Demystifying ATM/ADSL*, Wordware Publishing, Inc.
8. DSL Forum (2003), *ADSL2 and ADSL2plus – The New ADSL Standards*.
9. ANSI (2000), *High Bit Rate Digital Subscriber Line Second Generation (HDSL2)*, Standard TI. 418–2000.
10. ITU-T (2001), *Single-pair High Bit Rate Digital Subscriber Line (SHDSL) Transceivers*, Recommendation G.991.2.
11. ITU-T (1999), *Handshake Procedures for Digital Subscriber Line (DSL) Transceivers*, Recommendation G.994.1.
12. ANSI (2000), *Very High Bit Rate Digital Subscriber Line (VDSL)*, T1E1.4 committee document T1E1.4/2000-009R3 (LB941).
13. Azzam, A. and Ransom, N. (1999), *Broadband Access Technologies*, Mcgraw-Hill.
14. Lawrence, V. B., Smithwick, L.J., Werner, J.-J. and Zervos, N.A. (1996), Broadband Access to the Home on Copper, *Bell Labs Technical Journal*.
15. Starr, T., Sorbara, M., Cioffi, J.M. and Silverman, P.J. (2003), *DSL Advances*, Prentice Hall.

11

Fiber in the Last Mile

Ashwin Gumaste[1], Nasir Ghani[2] and Imrich Chlamtac[3]
[1]*Fujitsu Laboratories, Dallas, USA*
[2]*Tennessee Tech University, Cookeville, USA*
[3]*CreateNet Research Consortium, USA*

11.1 Introduction

The last mile represents the void in bandwidth surfacing from a high-speed core and the emergence of high-speed desktops at end user premises. Multiple technology solutions have been proposed to alleviate this bandwidth bottleneck between core networks and high-speed end users. The distributed nature of the last mile, combined with the bandwidth guarantees, have led to a situation whereby contemporary solutions such as DSL and wireless are either not cost effective or technology savvy enough to meet the emerging next generation high bandwidth needs. An effective bandwidth-oriented solution for the last mile needs a low deployment cost and good performance, using mature technology. Amongst the various physical layer solutions proposed, copper, wireless, power line and fiber are the most prominent for the last mile. The first three solutions are covered in different chapters in this book, while we focus on the optical-fiber-based solution for the last mile. Optical fiber, due to its low attenuation and high-bandwidth properties, is amongst the four physical media most suitable for last mile applications. However, the high end technological benefits also bring along a high deployment cost and a somewhat disruptive business case. Despite the seemingly high cost structure, fiber solutions are fast becoming prominent in the last mile for two reasons. Primarily, the other three technological solutions (copper, wireless and power line) cannot meet the emerging needs of high-bandwidth services like video-on-demand and collocated gaming, and hence fiber solutions are the best alternative for providing high-bandwidth ubiquitous services. Secondly, the cost of fiber solutions has considerably declined due to maturity in the associated technology, as well as the downturn affecting the telecommunication industry in general. With the telecommunication downturn setting in towards the latter half of 2000 and continuing through to 2003, the technologies that were used in the core were diverted in the access area – since the access

Broadband Services: Business Models and Technologies for Community Networks. Edited by I. Chlamtac,
A. Gumaste and C. Szabó © 2005 John Wiley & Sons, Ltd. ISBN 0-470-02248-5.

area was the only tangible business model at that time. This led to the last mile/access area being a plethora of technologies, and hence growth. Of the many fiber solutions proposed in the last mile, the passive optical network-based solution has been the most widely accepted so far. The passive optical network-based solution for the last mile has been technologically known for at least the last 15 years, even prior to the growth of fiber communication itself. The principle behind passive optical networking is to initiate and sustain communication between a central office and multiple end user sites. The last mile fiber network is subject to some constraints, such as lowering the equipment cost for end users, having some sort of optical multicast system (for video transmission) and guaranteeing enterprise level reliability in cases of node failure.

This chapter is organized as follows: in Section 11.2, we describe the topological model and architecture for PON systems. In Section 11.3, we will discuss the different types of passive optical network, namely APON, GPON, EPON and WDM PON. Subsequently, we will discuss components in PON systems, followed by a brief survey of the algorithms used to disseminate communication between central office and end users.

11.2 Topological and architectural model for PON systems: associated nomenclature

All known PON systems at this time are topologically configured as a star. The reason for choosing a star topology will be obvious at the end of the following discussion. The last mile represents a system whereby multiple end users have to be connected to a single reference point – a central office. The end users can be residential or enterprise customers, while the central office is a service provider point-of-presence (POP) site that terminates a high-bandwidth signal and distributes this into tributaries that can be fed to the end users. Hence, a star forms an almost ideal topology to connect the single central office to multiple end users. Also, as mentioned in the discussion before, the PON system has specific demands from a topology perspective. These demands include reliability, optical multicasting and low deployment cost. The star topology is ideal to meet these demands. Failure of a node in a star is unknown to other nodes in the network. When the central office sends a signal, the star network splits the signal into copies and hence broadcasts the signal to all the end users, leading to optical multicasting. Further, the system of multiple users communicating with a single central office also replicates somewhat a master–slave protocol model, whereby the slave nodes (end user equipment) need less complicated equipment than the master node, thus reducing deployment cost.

Consider the conceptual model in Figure 11.1. Shown here is a typical star network for the last mile built on passive optical network technology. The central office is connected on one side to the core network, while on the other side, it is connected to a fiber line that acts as a feed for multiple end user sites. The central office, due to its property of terminating core data and enticing a line of communication, is called the optical line terminal unit, or OLTU, or just OLT. On the other hand, the end user nodes are called optical network units, or ONUs. Communication from the OLT to the ONU is termed *downstream communication*, while that from the ONU to the OLT is termed *upstream communication*. The fiber line in principle allows bidirectional communication as long as the two modes of communication (upstream and downstream) are frequency (wavelength) separated. Typically, present PON systems use a higher order wavelength (1550 nm) for downstream communication, and a

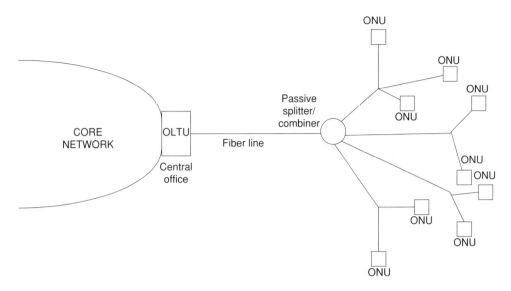

Figure 11.1 A passive optical network

lower order wavelength (1310 nm) for upstream communication. There is not much technical difference in the last mile as the two perform similarly, but the 1310 nm wavelength is chosen for upstream to save transmitter costs (a 1310 nm transmitter laser costs much less than a 1550 nm transmitter).

11.3 Types of passive optical network

The passive optical networking model shown in Figure 11.1 is a generic representation of the wide class of fiber networks for the last mile. There are multiple ways to facilitate communication between an OLT and ONUs in a PON. PON systems can then be classified by the technologies that they deploy. In principle, there are three broad classifications of PON systems:

1. ATM and broadband PONs.
2. Ethernet PONs.
3. WDM PONs.

All other PON nomenclature can be grouped into one of these types. Of course, there is also the lightframes framework that performs the same function as a PON, but this is not mentioned further here as it differs from a PON in operation [1].

11.3.1 ATM, or broadband PONs

The first category of PON system is the ATM PON. The ATM PON traces its roots to the FSAN (full-service access network) initiative by a select group 20 vendors and service providers. The FSAN group was first set up in 1995 and it submitted to the International Telecommunications

Union a document that was adopted by the ITU, to become what is commonly called the Broadband Optical Access System for PON, titled G.983.1.

In ATM PONs, at the OLT, the service provider terminates its connection in the form of several ATM and SONET lines. The OLT then filters out only the information streams destined for the users it supports. Typically, in ATM PONs, this filtering is done by virtual circuit/virtual path identifiers that are 28-bit addresses located in ATM cells. The OLT, in parallel, also informs ONUs about the VPI/VCI numbers that are relevant. This is done usually through the broadcast downstream medium, meaning that every ONU knows the VCI/VPI number associated with every other ONU in the network.

In the upstream, time division multiplexing is used for communication. Since many ONUs communicate with the OLT using the same channel, they all share the same bandwidth upstream. To avoid collision, a time division multiplexed protocol is initiated between the multiple ONUs, so that each ONU gets a fair share of the upstream channel. Apart from the TDM protocol, upstream communication also involves solution of the far–near problem, which is next described.

Since ONUs are at different distances, as compared to each other, the transmission delay between ONUs and OLT is the not the same for every ONU. Closer ONUs have less delay, while ONUs that are far away have a longer delay. To solve this problem, ONUs are synchronized by a technique called *ranging*. The ranging procedure allows the OLT to determine round trip delay times for every ONU, and hence allows creation of a guard band between TDM slots to 'scale' far and near nodes.

APONs come in three line speeds, 155 Mbps, 622 Mbps and 1.25 Gbps/2.48 Gbps, all being tried out separately. The first two line speeds come in symmetrical form, whereby the upstream and downstream speeds are nearly the same. In fact, the upstream line speed is slightly slower than the downstream one, and this is to induce extra provisioning and maintenance information. APONs allow up to 64 ONUs to be interfaced with a single OLT. However, by using native optical amplifiers, the number of ONUs can be increased to 128 or 256 with some degree of success. However, since the upstream channel is shared by all the ONUs, the scalability aspect of PON systems also depend on the service level agreements that are carried out. The only advantage APONs have over the other variants is in terms of management. Since service providers use ATM and SONET commonly in their backbones, APONs and BPONs form a homogeneous continuum in the management plane. Further, APONs and their more developed cousin – BPONs also have the added advantage of being able to provision relatively any wavelength granularity as desired. In contrast, TDM systems that existed in xDSL networks worked in integer multiplications of T1 lines. Here, the VCI could specify granularity completely reflecting the requirements of the service level agreement.

11.3.2 EPONs, or Ethernet PONs

The second implementation of a PON is that using Ethernet technology. Ethernet is the most prominent LAN technology and its extension to access is natural. However, adaptation of Ethernet in the access has had some basic problems, but despite these, it can be expected that Ethernet will play a rather important role in the PON segment in years to come. Native Ethernet frames are used for transmission in PONs and a protocol is implemented to arbitrate communication between the OLT and ONUs, as well as between multiple ONUs and the OLT.

The single biggest advantage Ethernet has over its ATM/BPON counterpart is its ability to do away with synchronization. Since Ethernet-based systems are not time synchronized, ONUs do not have to be synchronized with the OLT to establish communication. In the downstream, the OLT sends a continuous stream of Ethernet frames. Naturally, these are broadcast to all the ONUs. The frames have embedded within them a MAC destination address. Upon reception at the ONUs, the frames can either be discarded (if no match occurs), or processed for data (if a match does occur). In the upstream, however, the EPON system is slightly complicated. In the absence of a time division arbitration approach, as seen in A/BPONs, for the upstream, EPONs use some sort of a protocol that allows arbitration of communication between multiple ONUs.

Two examples of the protocol are shown in [2] and [3]. Multiple heuristic approaches can be chosen to design the protocol, and the existing EFM standardization document allows vendors to choose a protocol of their choice, as long as it conforms to certain specification requirements laid out by the EFM (Ethernet in the First Mile) group. In the upstream however, despite a protocol to arbitrate communication between multiple ONUs, the communication is still impromptu. This indicates that, in the absence of clear time division demarcations for upstream communication, the ONUs transmit quite randomly. This random upstream transmission requires a transmitter that can be fired ON and OFF relatively quickly. Likewise in the downstream, the frames being broadcast by the OLT may not always exist. This means that in the downstream, the OLT may just send data abruptly, and hence the ONUs have to receive this bursty form of transmission. This leads to a system where ONUs do not know when the data will come, nor do they have the phase and frequency of the modulated signal. Once the optical data arrives, the ONUs have to detect the light, convert it to electronic pulses and differentiate the zeros from the ones. This last procedure of differentiating the zeros from the ones is often done by correlating the detected bit stream with a locally generated clock. Of course, the locally generated clock has to match the bit stream generated upon detection, both in phase and frequency. Enabling these functions in the bursty (random in this case) system is done with what are known as *burst mode transceivers* and *burst mode receivers*. The design of burst mode transceivers and receivers will be discussed in subsequent sections.

11.3.3 WDM PONs

WDM means wavelength division multiplexing, and WDM communication was first introduced to allow the use of the massive bandwidth provided by optical fiber, by allotting a separate channel to every ONU. This is one of the most effective ways of utilizing the bandwidth provided by the fiber, but it is also extremely expensive to deploy. The cost increase comes from the need for optical components, such as WDM lasers, that are expensive. Further, WDM lasers are required to transmit only in the 1525–1565 nm band, known as the C band, and their transmission has to be temperature independent. The temperature independence is particularly important as lasers tend to drift in their emitted wavelength with changes in temperature. Since the ONU is placed in structures outdoors, such as curb sites or man holes, the temperatures tend to fluctuate. This appears to be a severe problem for WDM PONs. Hence, temperature independent lasers are designed using thermistors to control the laser driver current. A WDM PON is designed to have a single wavelength downstream from the OLT to all the ONUs, while in the upstream, each ONU has its own wavelength for transmission (Figure 11.2). This way, the issue of collision does not arise either. Each ONU can then transmit at any time and does

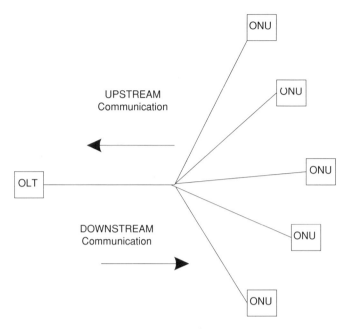

Figure 11.2 Upstream and downstream communication

not need to depend on the transmission timings of the other ONUs. However, each ONU still has to have either time synchronization with the OLT, or both OLT and ONU have to have burst mode receivers. The advantage of the WDM system is naturally the capacity it is able to get through the multiple channels it uses. On the negative front, the WDM PON system needs a large number of components. At the OLT, it needs filters to discriminate each upstream wavelength and feed the filtered wavelength to a receiver. If there are W ONUs, then there are W upstream wavelengths and the OLT then needs W receivers to detect these, in addition to a 1: W WDM filter. Further, at the ONU side, each ONU needs one laser corresponding to a particular transmission wavelength, from amongst the W possible wavelengths. This means that the number of laser types is W. Further, note that the number of wavelengths available is finite and this usually poses a severe limitation on the scalability of a WDM PON system. In summary, the WDM PON system is a good solution for a small number of high end customers (ONUs), but it has technological (laser drift), cost (laser) and scalability issues. Despite these, some vendors are proposing WDM PONs and variations as a possible fiber solution in the last mile.

11.4 Optical signal details to be considered while designing a PON system

So far we have considered the three basic types of PON system, and we can now investigate deeper the system level analysis of these systems. In this section we shall briefly consider the various optical issues that can affect PON systems. Primarily, optical transmission in fiber is limited by three well-defined phenomena: attenuation, dispersion and nonlinearities. Attenuation of an optical signal happens as a signal traverses through a fiber. Attenuation happens due to absorption, scattering and interference of light. Attenuation in a fiber is distinguished by its attenuation constant, alpha, that gives the loss of light in decibels per every traveled km (dB/km). Further, for the same fiber, the attenuation constant varies with wavelength. This is because absorption of light in silica varies with the wavelength of the traveling light. For the 1310 nm window used for upstream transmission, the constant is around 0.5 dB/km, while for the 1550 nm C band window of operation, the constant is around 0.2 dB/km for single mode fibers. The last term – single mode – determines the operation of a fiber. The fiber is here considered to be a cylindrical waveguide, and a single mode fiber allows just one dominant mode (analogous to the direction of communication) to exist. The attenuation factor gives rise to a loss budget, or power budget, in the PON system. This power budget determines the output power of the lasers at the OLT and ONU. The laser output power then depends on two factors, the attenuation in the system (length of fiber between OLT and ONU plus fiber splices) and the dynamic range of the photodetector.

The dynamic range of the photodetector has special significance on account of it being the single most important quantity that influences power budget in a system. The dynamic range gives the operating range of a photodetector (receiver). The photo diode at the receiver specifies two extreme power values: a lower value determining the minimum optical power it would need to function (this gives the minimum number of photons per bit required for detection) and an upper value determining the maximum input optical power that can be incident on the photo diode above which the photo diode will cease to work (the working reverse bias will break down). These two values, along with the attenuation, determine the power budget for the PON system.

Dispersion is the wavelength drift of a pulse in time. An optical pulse has multiple wavelengths embedded in it. These wavelengths, each travel at different velocities, called group velocities. Over time, the power in each wavelength travels a different distance, causing the pulse to lose shape. This phenomenon is called dispersion. Present PON systems are not affected by dispersion as transmission speeds are low – of the order of a single Gbps. But as transmission speeds increase, maybe to 10 Gbps and beyond, dispersion will be a key parameter to consider.

The optical fiber creates nonlinear impairments impeding the transmission of a signal through it. Though these effects are negligible in PON systems, as the speed of transmission increases, these effects are likely to affect transmission. The effects are: self-phase modulation, cross-phase modulation and four-wave mixing.

11.4.1 Optical signal-to-noise ratio

The optical signal-to-noise ratio is the single most important quantity when describing the optical signal in a transmission system. The optical signal-to-noise ratio, or OSNR, gives the ratio of the optical signal power to the noise content. It is a statistical value, often calculated deterministically. The OSNR determines the actual useful content of a signal. In PON systems, the OSNR is only somewhat important as there are no optical amplifiers (the chief source for noise). However, OSNR-based designs are equally valid for power budget analysis of PON systems.

11.5 Components for PON systems

Generically, the PON system consists of the OLT and the ONU. We shall, in this section, describe the subsystem level configuration of the OLT and the ONU. Amongst the three types of PON system described earlier, we will keep the OLT and ONU discussion to a generic level, signifying that the following discussion is valid for APONs, EPONs and WDM PONs, with some minor modifications.

Shown in Figure 11.3 is the subsystem level layout of the OLT. As can be seen, the OLT is a 2×2 port device, in the sense that it transmits and receives data from a carrier termination point, as well as transmiting and receiving data to multiple ONUs.

The OLT takes in information frames from the service provider feed (SP feed) and further deciphers these frames. It may either reframe the information that is to be transmitted to the ONUs, or simply tag the information frames with some protocol identification information. The information from the SP feed is correlated to the protocol stack, so that downstream broadcast and upstream communication are linked. The relaying of downstream communication is buffered, and flow control is carried out through buffers in the protocol stack. The buffers then control the flow of information. The buffers also add control information to the downstream frames. This control information is critical to arbitrate communication in the upstream. This is primarily done by the control information being piggybacked on the downstream frames to all the ONUs. The ONUs can then transmit in the upstream, depending on the information that they receive. Hence, this way, the OLT controls upstream transmission through a piggyback procedure. The information frames plus piggybacked control signals are then broadcast downstream to the ONUs, through a burst mode transmitter. For cases where the OLT uses a synchronous protocol, as in the case of a BPON system, etc., the transmitter need not be burst mode but can be a simple direct modulated laser diode.

In the other direction of communication (the upstream), the OLT receives optical signals from the ONUs. The OLT receives the upstream frames through a burst mode receiver. The burst mode receiver is critical, irrespective of the communication methodology, as the upstream is often bursty in data flow.

Figure 11.4 shows the ONU subsystem layout. The ONU has a single fiber interface connected to the network side, to the OLT. It also may have multiple local interfaces to connect to multiple physical devices at customer premises. If we envision an ONU located at a curb site (in the case of FTTC – fiber-to-the-curb), then the ONU can be connected to end users via wireless, copper or even multimode fiber lines, depending on the system needs. In the downstream, the signal from the OLT enters the ONU and is filtered optically to ensure upstream and downstream communication is separate. The optical signal is then converted to electric bit streams through a burst mode receiver and clock recovery circuit. The burst mode receiver here is the most critical aspect of the ONU. Since the downstream data is not continuous, a normal detector cannot detect the randomly varying stream of information properly. This is because, for a typical detector to detect the information frames it needs to know the amplitude and frequency/phase of the incoming signal beforehand. However, if the data is not continuous, as in the case of downstream transmission, a normal photodetector will soon lose lock with the incoming bit stream, and hence generate garbled frames. To avoid this, a burst mode receiver is used. The burst mode receiver is able to detect information frames without knowing the amplitude or phase/frequency of the incoming signal. There are multiple technologies known for burst mode receivers. The most common, however, is to decipher the amplitude and phase information through an embedded string of bits within the preamble of an incoming frame. Other methods of deciphering the amplitude and phase are by using nonlinear elements that act as square law loops, enabling multiple harmonic components of the frequency and phase to be generated.

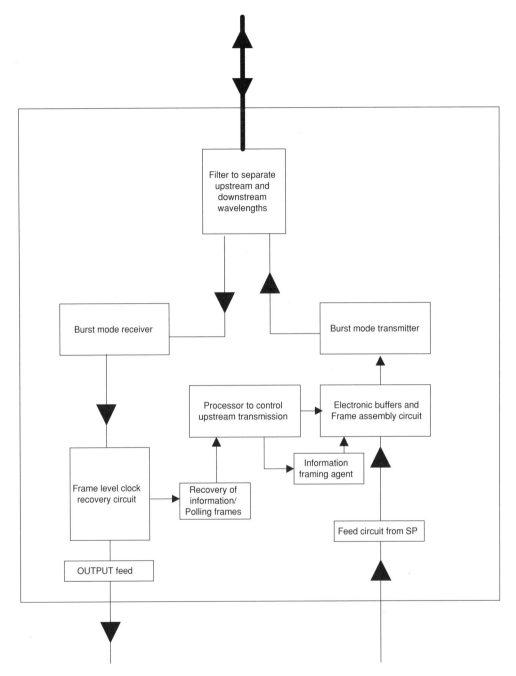

Figure 11.3   OLT subsystem layout

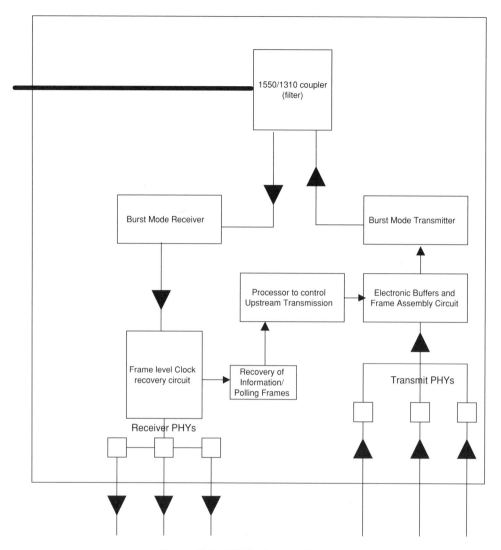

Figure 11.4   ONU subsystem layout

Once the bit stream is converted into an electronic data stream, frames are recovered and data information is separated from the associated control information. This control information is the information that the OLT sends to the ONU by piggybacking on the downstream frames. Upon recovery of this control information at the ONU, it is then sent to a local processor. The local processor correlates this information and gates the upstream data, as well as locally controlling the transmitter (for upstream communication) based on the control information. The data frames recovered from the burst mode receiver are then sent to respective PHYs (physical interfaces) for transmission to the CPE (customer premises equipment). The ONU also accepts data from CPEs and has the responsibility of transmitting this data upstream to the OLT. It takes in frames from the CPE and stores them locally in the buffer. Depending on the transmitting rights of the upstream channel, the ONU then controls the flow of the frames through the use of these buffers. The frames are transmitted to the OLT by the burst mode transceiver (transmitter). The burst mode transceiver, like the burst mode receiver, is able to transmit frames randomly without emitting continuous light – hence the term burst mode transceiver.

11.6 Protocol requirements for OLT–ONU interaction

Based on the component description in the previous section, we will now discuss the protocol that works between the ONU and the OLT for transmission. As evident from the earlier discussion, the downstream communication is a simple case of broadcast communication and a protocol is not of much importance in any kind of arbitration. It is the upstream communication that requires a protocol for arbitration. In the upstream, there are multiple ONUs trying to transmit to a single OLT via a combiner and passive channel. It is therefore in the upstream that there is a need for a protocol to arbitrate communication between the multiple ONUs, so that the ONUs do not transmit all at once, or any two at once, resulting in collision. It is the collision avoidance mechanism that is the primary motivation of the protocol. Secondly, the protocol does need to ensure fair transmission rights of the channel (upstream) to carry out specific service level agreements (SLAs) between ONUs and OLT. There are multiple ways to implement the upstream protocol and there is no specific standard for this upstream protocol.

Generically, the upstream protocol is multipoint to single point in transmission nature, and hence the EFM standard calls it the MPCP, or Multipoint Control Protocol. The implementation of this protocol, however, is vendor-specific and we shall now discuss two academically known protocols.

11.6.1 Transmission upon reception (TUR)

First proposed in [2], this protocol was designed to arbitrate communication between ONUs and OLT, with a particular interest in supporting dynamic bandwidth allocation. The protocol is designed for EPON systems only. The protocol is able to support dynamic allocation of bandwidth to meet emerging bandwidth-on-demand applications. The protocol, as the name suggests, controls upstream transmission, based (proportionally) on downstream allocation in time (Figure 11.5). In the protocol, an ONU gets transmission rights of the upstream channel, proportional to the amount of data it receives. This means that if an ONU received data from time t_1 to t_2, then it would be able to transmit from some time t_k to $t_k + (t_2 - t_1)$. In this case, t_k signifies the earliest time when the upstream will be free. This way, the upstream is utilized quite effectively without a need for individual nodes to be synchronized. TUR also allows SLAs to be met by sending dummy downstream frames, and thereby granting a particular ONU access to the upstream channel, even when it actually does not have data destined for itself. From an implementation perspective, TUR has a good combination of distributed intelligence, whereby both OLT and ONU have some processing.

11.6.2 IPACT: interleaved polling with adaptive cycle time

The IPACT was first introduced in [3] and it basically is one of the first implementations of a PON protocol. Like TUR (introduced much later), IPACT is also primarily suited for EPON

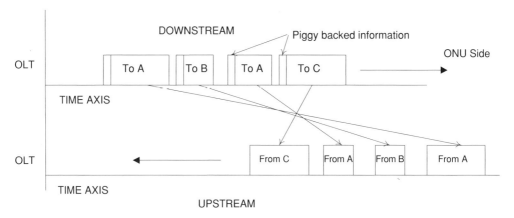

Figure 11.5   Protocol dissemination between OLT and ONU: TUR (transmission upon reception) [2]

systems. IPACT is a polling protocol, in the sense that an OLT proactively tells an ONU when to transmit. The OLT is assumed to know the buffer state of each ONU, and the OLT then computes which ONU should be next scheduled for transmission. The protocol assumes that the OLT knows the round trip transmission time (RTT) and hence is able to precisely tell each ONU when to transmit. Some degree of synchronization is needed here, as the processing is done by just the OLT, the ONUs act as mere slaves, following somewhat strictly the OLT commands.

11.7 MPCP classifications and requirements

In this section we shall discuss the salient features that a good MPCP will provide to a PON system.

11.7.1 Dynamic provisioning

An MPCP should be able to dynamically provision services to ONUs on a real-time basis. The dynamic nature of services often means that the service guarantees have to change with time. The MPCP should be able to arbitrate bandwidth between multiple ONUs dynamically. Further, optimal means of bandwidth arbitration are needed in order for the ONUs to efficiently utilize the upstream bandwidth.

11.7.2 Fairness

Since a large number of ONUs use the same upstream channel, fairness is a virtue that is essential for revenue bearing services. Polling protocols are able to guarantee fairness to ONUs quite well. Fairness here is defined as the ratio of the net amount of time an ONU gets upstream transmission of the channel to the total time. If there are K ONUs in a system, and the upstream line rate is C bits/s, then, in a fair system, each ONU should be able to service customers connected to it with a line speed of C/K bits/s without dropping any frames.

11.7.3 QoS

Apart from the generic fairness guarantee, a PON protocol must also be able to meet any QoS needs of the ONUs. Specifically, QoS means that fairness is not totally achieved. QoS needs may vary from bit rate specific to latency specific. Bit-rate-specific service guarantees are easier to provision, while latency specific often means complex algorithms for dropping packets and managing queues. Common algorithms like weighted fairness queues and random early detection are used for QoS provisioning.

11.7.4 Stability and convergence

PON protocols have to be stable, failing which the system would have high packet loss. Stability is determined by the net drift between OLT and ONU. Convergence of this drift into manageable margins is a prerequisite of the PON system, and this convergence should happen relatively quickly. The higher the time required for convergence, the greater the frame loss probability.

11.7.5 SLA provisioning

Like QoS guarantee, service level agreement (SLA) provisioning is crucial to bringing in revenue. SLA provisioning is another prerequisite of the PON protocol. There are two aspects of the SLA that need attention. First, the management aspect, whereby the ONU informs the OLT of the SLA and the OLT determines if the SLA is valid (secure identification, etc.). Secondly, the OLT has to incorporate the new SLA into the protocol, and this procedure should be done in real time.

11.8 Summary

In this chapter we discussed fiber technologies in the last mile. Fiber in the last mile is popularly provisioned through the concept of passive optical networks, or PONs. PON technologies come in three flavors, APONs/BPONs, EPONs and WDM PONs. There are various schools of thought advocating each different technology. While A/BPONs are TDM-based technologies, EPONs provide a packet-oriented scheme. WDM PONs, on the other hand, can deploy any of the two other technologies. A/BPONs are more suited for voice communication, while EPONs are more suited for data-centric communication. A/BPONs, due to the synchronization requirement, are more expensive than EPONs in terms of technology. However, EPONs have some issues with provisioning strict SLAs.

In summary, PON systems are likely to be dominant in the last mile for fiber communication in years to come. Though the star topology is clearly defined for these kinds of system, it would be worth investigating other topologies like mesh and fiber in the local loop. Since network utilization in the access area is low, packet-based systems can be intuitively considered as a good technology and business alternative. It is our belief that packet networks for fiber in the last mile using meshed topologies would be a new horizon in research for this area of communication.

References

1. Gumaste, A. (2004), *First Mile Access Networks and Enabling Technologies*, Macmillan Publishers and Pearson Education.
2. Gumaste, A. and Chlamtac, I. (2003), A Protocol to Implement Ethernet Passive Optical Networks (EPON), in *Proceedings of IEEE International Conference on Communications*, ICC 2003, Anchorage, Alaska. May 2003.
3. Kramer, G. and Mukherjee, B. (2002), IPACT: A Dynamic Protocol for an Ethernet PON (EPON), *IEEE Communications Magazine*, February.

12

Ethernet in the First Mile

Wael William Diab

IEEE 802.3ah, IEEE 802.3 (Ethernet) and Cisco Systems, San Jose, USA

12.1 Introduction

Ethernet in the first mile, frequently called EFM, was a standardization effort formally called P802.3ah, a project under the IEEE 802 LMSC Committee and part of the 802.3 Working Group.

This chapter will start by introducing the IEEE process and giving some background, with a timeline on the EFM project. Next, an overview of each set of technologies covered by this emerging standard is developed. The chapter concludes with a series of discussion topics on the architectures standardized and the technology choices.

12.2 Overview of the IEEE, the process and the 802.3 Working Group

12.2.1 The LMSC

The EFM project is part of the IEEE 802 LMSC Committee. This is the IEEE 802 LAN, MAN Standards Committee (LMSC) and it functions as the sponsor to a number of projects originating in its Working Groups. Structurally, the LMSC consists of the Sponsor Executive Committee, SEC 802.0, and Working Groups and Technical Advisory Groups. Some of the active Working Groups and Technical Advisory Groups include 802.1 Higher layer protocols, 802.11 Wireless LAN, 802.17 RPR and 802.3 CSMA/CD (Ethernet). Historically, the group's first meeting was in February of 1980. It was originally called the Local Network Standards Committee. The MAN scope was added later. The number, 802, which is very popular today because of the tremendous success of projects within wireless, Ethernet and others, was simply the next project number of IEEE at the time.

Broadband Services: Business Models and Technologies for Community Networks. Edited by I. Chlamtac, A. Gumaste and C. Szabó © 2005 John Wiley & Sons, Ltd. ISBN 0-470-02248-5.

12.2.2 The Ethernet Working Group and the IEEE process

One of the older and more successful Working Groups in the LMSC is 802.3, formally known as CSMA/CD but more commonly referred to as Ethernet. Amongst the number of successful projects standardized by 802.3 are 802.3ae 10 Gigabit Ethernet, 802.3ad Link Aggregation and 802.3af Power over Ethernet, which have been standardized recently, and 802.3z Gigabit Ethernet, 802.3 100BASE-TX and 10BASE-T, which are amongst some of the more popular projects that were standardized earlier.

Typically, a new project goes through the following phases:

1. Call for interest to form a Study Group. Usually a few companies or parties interested in standardizing an emerging technology put forth a presentation and proposal to the Working Group. If 50 % of the Working Group votes in favor, a Study Group is formed to look at the new space.
2. PAR and five Criteria to form a Task Force. The Study Group's main focus is to produce a Project Authorization Requirement (PAR), which includes a set of objects that meet the five criteria (see discussion below). Once approved, the Study Group officially becomes a Task Force of the 802.3 Working Group.
3. Baseline proposals. The Task Force's first major hurdle is to accept a set of baseline proposals that meet the agreed objectives. Usually this is a set of Powerpoint slides that are often endorsed by a number of the participating member companies. As with all technical decisions, a 75 % consensus is required.
4. Creation of the document and Task Force Review. The baseline documents are then transcribed to a working document that is reviewed by the Task Force for technical completeness and maintained by a team of elected technical editors.
5. Working Group Ballot of the document. Once the Task Force is done reviewing the document, a more formal review in the form of a Working Group Ballot is conducted. This ensures that the five criteria and objectives are met and the scope of the voters spans all of 802.3 to ensure that Ethernet members from other Task Forces get the chance to review the document.
6. LMSC Ballot of the document. The final review is sponsored by the LMSC itself, and the resulting ballot group that is formed is extremely diverse, drawing on other members of 802 to review the work. Again, a 75 % approval is required at this stage.
7. Formal approval by the Standards Board and REVCOM. Once all balloting is complete, the document is submitted for approval by the standards board and REVCOM. This initiates the official publication of the standard.

12.2.3 A brief discussion on the five criteria

The historic success of Ethernet, in part, has been driven by its process-driven approach to creating standards. Key amongst which have been the five criteria that are an essential part of getting a project started and authorized. Below is a list of the five criteria, which have often been called the 'five critters' by some of the veteran participants, as they really span the entire scope of a new technology:

1. *Technical feasibility.* Is this new technology technically feasible in the timeframe of this project?

2. *Economic feasibility*. Is this new technology economically feasible in the timeframe of this project?
3. *Distinct identity*. Is this new technology different from existing standards?
4. *Compatibility*. This looks at compatibility with existing standards. Given the rich history of 802.3, this is particularly important to new projects in Ethernet.
5. *Broad market potential*. When all is said and done, how much market potential and traction will this new technology achieve?

A new project has to meet all five criteria to go forward, and as you can see, this requires a great deal of thought and data collection before any serious commitment is given to the project. Moreover, the individual objectives that are agreed to within the project have to meet these as well.

12.3 Overview of the EFM Task Force and timeline

12.3.1 P802.3ah: overview of EFM

Ethernet has enjoyed great success in the LAN (local area network) market. Moreover, while some of its technologies have been used in the MAN space, by and large, Ethernet had not addressed that market in the past. Thus, it should not have been a great surprise when the call for interest drew an unusually high turn out. More importantly, the diversity of the members that have contributed to standardizing the access space has been wide, in scope and expertise.

The objectives in EFM include requirements for new copper and optical physical layer devices. Moreover, the requirements also include new optical physical layer devices, logic enhancements to allow for point-to-multipoint operation and physical layer independent Ethernet operations, administration and maintenance (EFM OAM). Consequently, the Task Force was organized along those lines: four tracks, P2MP Logic, Optics, Copper and OAM, with each track having a chair, chief technical editor and a team of technical editors, as well as participating industry experts.

12.3.2 Practically, why is 802.3ah important?

There are a number of reasons, some of which have already been mentioned, as to why EFM has been important and successful in drawing a lot of industry attention. Below are several:

- it leverages a large installed base of Ethernet ports;
- it eliminates unnecessary protocol conversion;
- the architecture choices, described below, allow for future low-cost and high-volume implementations;
- the architecture choices allow for high-speed optical, as well as higher speed DSL, copper links;
- the architecture allows for migration from copper to optical via hybrid implementations;
- it does not limit itself to, or attempt to define a scope for, certain applications, but rather enables a host of applications that include voice, video and data;

- contributing participants included a host of industry representatives, the diversity of which allows the standard to address a diversity of issues:
  - customers: service providers like ILECs and CLECs;
  - vendors: system, silicon and optical vendors.

12.3.3 Timeline of the EFM project

The call for interest for EFM happened in November of 2000. Subsequently, an initial document was created in May of 2002, a Working Group Ballot was initiated in July of 2003 and an LMSC Sponsor Ballot was initiated in January of 2004. At the time of writing, the EFM document is winding down through the LMSC Ballot and has been targeted for a June 2004 approval by the standards board.

12.4 Overview of OAM

12.4.1 High-level architecture

OAM stands for operations, administration and maintenance. The technology was intended for monitoring link operation, monitoring link health and improving fault isolation. The capabilities listed are intended to drive down the operation expense of a service provider that deploys EFM technologies in its access space.

There are three important high-level characteristics of EFM's OAM:

1. It is an optional feature.
2. It has a frame-based architecture that may be implemented in hardware or software.
3. It is restricted to the actual access link, i.e. does not cover end-to-end OAM.

It is noteworthy to mention that the tremendous interest in OAM prompted modifications of the exiting 10 Gigabit Optical Ethernet standard to allow for OAM operation.

Architecturally, the OAM sublayer sits on top of the MAC control sublayer in the OSI layer stack (Figure 12.1). This is towards the top of the data link layer (layer two).

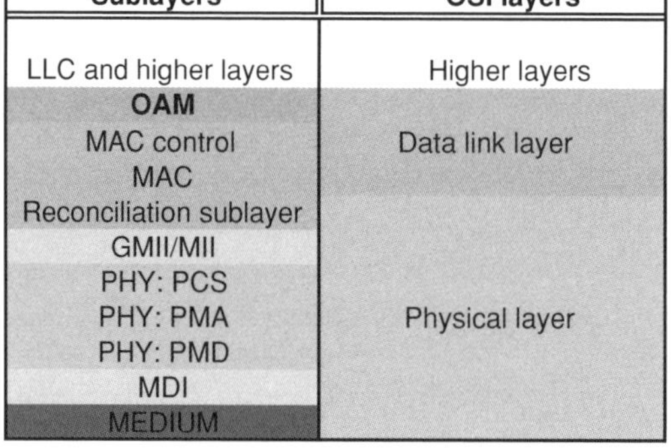

Sublayers	OSI layers
LLC and higher layers	Higher layers
OAM	Data link layer
MAC control	
MAC	
Reconciliation sublayer	
GMII/MII	Physical layer
PHY: PCS	
PHY: PMA	
PHY: PMD	
MDI	
MEDIUM	

Figure 12.1 The position of OAM within the protocol stack

Fundamentally, the OAM architecture relies on the slow frame transmission protocol, as defined by Annex 43B of 802.3. There are two important modifications that EFM makes to 43B to accommodate specific OAM requirements:

1. The frame size recommendation of 128 octets does not apply.
2. The maximum speed of transmission is 10 frames per second, as opposed to the previously defined 5 frames per second.

The OAM frames are referred to as OAMPDUs. This stands for operations, administration and maintenance protocol data units. These frames must have standard Ethernet frame lengths of 64 bytes to 1518 bytes and must be untagged. Moreover, they use a protocol subport type of 3, in contrast to previous uses of this protocol, such as number 2 for link aggregation.

OAM is supported on all point-to-point and emulated point-to-point links. It includes a discovery mechanism to allow two stations to discover if the other end is supporting this feature. This is due to the optional nature of OAM on EFM links.

12.4.2 Functionality

OAM functionality falls under three broad areas: remote fault indication, remote loop back and link monitoring.

Remote fault indication is a mechanism that is used to indicate to a transmitting link partner that the receive path of the current link is faulty. This provides powerful diagnostic information, since under legacy link setups the link would simply go down, leaving the IT professional with little indication as to which side is faulty. As discussed below, there is some restriction on the use of Gigabit auto negotiation with this feature.

Remote loop back, on the other hand, provides the IT professional with a diagnostic tool. The tool allows the local station to set the remote station in loop back *at the level of the OAM sublayer*. Frames sent from the local station after this operation is done are looped back to the local station. Apart from checking the integrity of the path, other statistics like BER (bit error rate) can be measured with a relative degree of accuracy.

The final area covered by OAM is link monitoring. This can be viewed as two separate subfunctions. The first allows the local station to poll (but not set) the variables defined by the management MIB and implemented in the remote device (much of the management is optional) as defined by Clause 30.

The second subarea is the ability to communicate critical events such as dying gasp or an unspecified critical event. In either case, this information would be critical in an access topology where the IT professional's physical access to the remote terminal may be limited or expensive (to deploy a field technician to the remote site).

12.4.3 Single link management

As mentioned above, the functionality defined by EFM is intended to manage a single link, not end-to-end. Consequently, it is important that the frames discussed above do not traverse other links and are thus not forwarded MAC clients, such as bridges.

12.4.4 Restrictions

Many IT professionals are familiar with 802.3 features that are typically deployed in networks. Due to the management nature of OAM, a couple of features are mutually exclusive with OAM. Specifically, flow control and the autonegotiation feature of Gigabit Ethernet.

Flow control, commonly, is a capability that relies on a special MAC PAUSE frame that is defined by Annex 31B. This mechanism is used to throttle back traffic and can inhibit all frames, including the OAMPDUs. Thus, care should be taken when enabling both OAM and PAUSE.

OAM fault signaling will not work with the Gigabit Ethernet autonegotiation state machine, as that requires the link to be up in both directions, which is exactly the signaling scenario in OAM. Thus, the Gigabit Ethernet autonegotiation must be disabled.

The standard summarizes some areas that are not covered by the EFM OAM. Some have been previously mentioned in this overview, nonetheless, since many terms are often associated with the term OAM, a complete list is given below:

- management functions, such as station management and protection switching;
- provisioning issues, including bandwidth allocation, rate, speed, etc.;
- privacy and security areas;
- ability to remotely set (write) the remote management interface;
- end-to-end, scope limited to the link.

12.5 Overview of copper

12.5.1 High-level architecture

EFM defines two point-to-point copper technologies, often referred to as the DSL technologies. These are 2BASE-TL and 10PASS-TS. Due to the varying transmission nature of the underlying copper medium (associated with legacy telephony), the actual rate and reach of a particular point-to-point copper link will depend on actual characteristics of the medium they are attached to.

Nonetheless, the 2 in the 2BASE-TL is intended to reference a nominal bit rate of 2 Mbps at a nominal reach of 2700 meters. Similarly, the 10 in 10PASS-TS is for 10 Mbps at 750 meters. This assumes that a single nonloaded copper pair is used.

Unlike the OAM functionality described above, the two PHYs defined by copper attach to a simplified full duplex MAC, as defined by Annex 4A, via the traditional MII interface. Each PHY is then divided into four sublayers: PCS (physical coding sublayer), TC (transmission convergence), PMA (physical medium attachment) and PMD (physical medium dependent). With the exception of the TC sublayer, the remaining sublayers are consistent with previously defined PHYs in IEEE.

The architecture mirrors and references other xDSL architectures, as defined by ATIS T1, ETSI and ITU-T standards. To that effect, the gamma and alpha/beta interfaces are also defined to mirror the other documents. Specifically, the gamma interface sits between the PCS and TC as outlined below. Moreover, the alpha/beta interface sits between the TC and the PMA.

For each of the technologies, 2BASE-TL and 10PASS-TS, two ends of the link are designated. One is named the CO, central office, and the other the CPE, customer premises equipment. If the equipment is used in an enterprise environment, then the IT professional can arbitrarily designate one end to be the CO and the other the CPE.

The PCS and TC sublayers are part of a generic architecture defined for both technologies in Clause 61 of the EFM document. The remaining layers underneath the alpha/beta interface are specific to each technology type.

The astute reader will note that the MII connects to a MAC and that subsystem operates at 100 Mbps, while the bulk of the PHY subsystem operates at varying, and predominantly lower, rates. Indeed, this provides a need for a rate matching mechanism that is implemented within the PCS layer.

12.5.2 More on the specifics of each type of copper link

10PASS-TS is intended for shorter reach but higher speed links. To achieve full duplex on a single pair, a frequency division duplexing technique is used. The signaling system used references the VDSL signaling system, defined by ANSI T1.424 using DMT, discrete multitone, modulation.

Unlike the passband signaling used by the higher speed technology, 2BASE-TL is intended for further reach and lower rate applications. This baseband signaling system references SHDSL, Single-pair High-speed Digital Subscriber Line, defined by the ITU.

12.5.3 Support for multiple pairs

Multiple copper pair performance is supported as an optional feature to allow for further increases in bandwidth where required. This function is implemented above the alpha/beta interface.

12.6 Overview of optics

12.6.1 High-level summary

Broadly speaking, the optics group defines four sets of solutions: point-to-point 100 M solutions, point-to-point Gigabit solutions, point-to-multipoint solutions and extended temperature optics. Each of these areas contains multiple different solutions within it.

The reason for the multitude of areas and solutions within optics was to consolidate the optics expertise within one room. This resulted in a compatible and consistent writing of all the optics requirements, solutions and testing.

Point-to-point, P2P, optics are used over a dedicated single link, whereby there is one device on either end of the link. In the access space, this is usually a CO and a CPE, similar to the copper architecture.

Point-to-multipoint, P2MP, optics are intended to reduce some of the optical and fiber costs by sharing both amongst multiple ports. By way of example, a 15-subscriber network would have 15 pairs (one on each end) of optics for a total of 30 optics. In a P2MP topology, only one optic is required at the CO end and 15 at the subscriber end, thus reducing the total number of optics in the system to 16.

The P2MP solution introduces a new protocol, discussed in the next overview section, as well as additional requirements on the optics. The P2MP solution defined by EFM is often referred to as an EPON, which stands for Ethernet passive optical network. The term passive is motivated by the idea that the fiber is shared, since only one strand can go to the single CO optic. This requires a passive optical splitter to split the fiber out to each individual end node.

Both families of P2MP optical solutions defined allow up to 16 end stations to be connected on a single PON.

12.6.2 High-level architecture: common motivations

The optics section simply defines the PMD portion of each PHY technology specified. Below is a high-level list of all the various PMDs. Note that the bidirectional technologies, as well as the P2MP technologies, have unique devices on either end of the link, and hence the added qualifier for each name. For the other technologies, each of the devices at the end of the link are symmetric and use the exact same PMD. All technologies are full duplex.

- Point-to-point 100 M solutions (P2P)
  - 100BASE-LX10: 10 km dual single-mode fiber solution;
  - 100BASE-BX10: 10 km dual wavelength dual single-mode fiber solution (bidirectional)
    - 100BASE-BX10-U;
    - 100BASE-BX10-D.
- Point-to-point 1 Gigabit solutions (P2P)
  - 1000BASE-LX10: 10 km dual single-mode fiber solution;
    - allows operation over short reach multimode fiber as well.
  - 1000BASE-BX10: 10 km dual single-mode wavelength dual fiber solution (bidirectional);
    - 1000BASE-BX10-U;
    - 1000BASE-BX10-D.
- Point-to-multipoint solutions (P2MP)
  - 1000BASE-PX10: 1:16, 10 km single single-mode fiber;
    - 1000BASE-PX10-U;
    - 1000BASE-PX10-D.
  - 1000BASE-PX20: 1:16, 20 km single single-mode fiber;
    - 1000BASE-PX20-U;
    - 1000BASE-PX20-D.
- Extended temperature optics
  - no defined range;
  - low range;
  - high range;
  - both high and low range.

12.6.3 Detailed description of the optical devices

Table 12.1 describes the wavelength, distance supported, fiber supported and technology family of each of the optical PMDs defined by EFM.

Table 12.1 The optical PMDs, as defined by EFM

PMD name	Technology family	Direction	Nominal wavelength	Distance	Fiber characteristics
100BASE-LX10	P2P 100 Mb/s	Symmetric	1310	10 km	B1.1, B1.3 SMF
100BASE-BX10-D	P2P 100 Mb/s	Downstream	1550	10 km	B1.1, B1.3 SMF
100BASE-BX10-U	P2P 100 Mb/s	Upstream	1310	10 km	B1.1, B1.3 SMF
1000BASE-LX10	P2P 1000 Mb/s	Symmetric	1310	10 km; 0.55 km	B1.1, B1.3 SMF, 50 μm and 62.5 μm MMF
1000BASE-BX10-D	P2P 1000 Mb/s	Downstream	1490	10 km	B1.1, B1.3 SMF
1000BASE-BX10-U	P2P 1000 Mb/s	Upstream	1310	10 km	B1.1, B1.3 SMF
1000BASE-PX10-D	P2MP (EPON)1000 Mb/s	Downstream	1490	10 km	B1.1, B1.3 SMF
1000BASE-PX10-U	P2MP (EPON)1000 Mb/s	Upstream	1310	10 km	B1.1, B1.3 SMF
1000BASE-PX20-D	P2MP (EPON)1000 Mb/s	Downstream	1490	20 km	B1.1, B1.3 SMF
1000BASE-PX20-U	P2MP (EPON)1000 Mb/s	Upstream	1310	20 km	B1.1, B1.3 SMF

Notes: P2P = Point-to-point; P2MP = Point-to-multipoint; SMF = Single mode fiber;
MMF = Multimode fiber

12.6.4 Philosophy for both single and dual fiber 100 M

The 100 M objective was a late addition to the standard but enjoyed a huge amount of support.
The main motivation was driven by two philosophies. First was to provide low-cost fiber access
solutions. Second was to standardize the solutions given the myriad of variations and existing
solutions within the market for this lower speed optical application.

To understand the rationale behind not defining anything new above and beyond the PMD,
one has to understand the cost structure behind the 100 M solutions. First, 100 M silicon already
exists for optical transmission over multimode fiber (100BASE-FX). Second, the volume of
the 100BASE-FX ports is, in large part, driven by the volume of the 100BASE-TX copper
parts, as very often the same piece of silicon is used for both applications. Third, there are a
number of transmit and receive specifications for both the single single-mode fiber and dual
single-mode fiber applications. Thus, it was important to pick a combination that would yield
the lowest costs in the long run.

A similar philosophy was followed for the Gigabit point-to-point solutions: no silicon was
modified. However, there was a more limited number of existing optical solutions at a Gigabit
than 100 M.

12.6.5 Philosophy for dual wavelength for the single fiber solutions

As was detailed in the previous table, all the optical technologies over single-mode fiber utilize
a dual wavelength solution, as opposed to a single wavelength solution. Much discussion and
debate was spent on this. This section presents the advantages and disadvantages of each, and
describes why the dual wavelength route was taken.

The primary advantage of a single wavelength system is that it utilizes the same part on
either end of the link, so the entries in the above table would have been a single symmetric
entry for the bidirectional links. The consequences of this are higher volume (double), which

arguably would drive down the cost. Also, a single and symmetric part would potentially make inventory and management on the service provider's part easier.

Moreover, in the case of the lower 100 M speed, a Fabry–Perot (FP) laser was feasible for both the 1310 and the 1550 wavelength lasers, due to the limited dispersion. For the higher Gigabit speeds, the longer wavelength of 1490 required a more expensive laser, a DFB, to accommodate the dispersion issues.

The biggest issue with the single wavelength system, however, is the technical issues that could show up during operation. Specifically, a single wavelength system is considerably more sensitive to reflections on the line. While such reflections can easily be overcome in a lab environment, in an operational environment, issues like air gaps and the number of connectors in the line affect the reflectance. Furthermore, most system engineers and IT professionals are trained to think in terms of pure loss on a link, which is associated with the number of connectors and length of fiber. The reflectance issue introduces an additional dimension of thinking of the reflections that occur at each interface that, for the users, is an additional complexity that they are not used to thinking about, or equipped to measure.

Finally, an additional complexity with the single wavelength system is the need to better isolate the transmitter from the receiver, as there are cross talk, as well as second order, effects that can cause the transmitted light to feed into the receiver and vice versa. This requires an isolator, which could put the ultimate cost of the system at a comparable one with the lower volume, but isolator-less, dual wavelength system.

12.6.6 Differences between the 100 M and 1 Gigabit point-to-point solutions

A few key differences have been mentioned, such as the need for a more costly laser at the higher wavelength (downstream direction) for Gigabit as opposed to 100 M on the bidirectional systems. This section describes some other key differences for point-to-point systems.

In the case of the dual fiber systems, the Gigabit solution was designed to operate over multimode systems as well, but at a much shorter reach. This was mainly due to the historic 1000BASE-LX that was defined by the Gigabit Ethernet standard (802.3z) and was specified to do both. The idea was that 1000BASE-LX10 would replace that for both single and multimode applications.

In the case of the single fiber systems, the Gigabit solution was designed to operate at a slightly lower wavelength of 1490 in the downstream (as opposed to the 1550 for the 100 M). The motivation was to stay away from the C Band, which many vendors and customers indicated could be of use for an analog overlay.

The astute reader will note that the same rationale could have been applied for the lower speeds. Indeed, he/she would be right. However, there was a motivation to harmonize EFM with TTC, an existing Japanese standard for dual wavelength operation at 100 M, to allow for a broader use of the optical parts that would hopefully drive volume up and cost down.

It is noteworthy to mention that while the lower wavelength of 1490 does not preclude an overlay, it does not specify it either. Parts that do implement the overlay may require additional components.

12.6.7 New requirements for EPONs

As mentioned above, EPONs allow a saving on the number of optics needed when multiple end stations are involved. Moreover, it was eluded to that there are a number of additional requirements on these optics. Broadly, there are two areas that are different.

First, the cable specifications and network topology definitions are more complex. With a P2P solution, there is one topology to connect the two devices and the specifications or variability are centered on cable loss. To the first order, this is predominantly the number of connectors and length of the fiber. In a P2MP topology, there is an endless number of ways that a set of end stations can be connected to the CO. For instance, the fiber from the CO optic can be split at once into the required number of fibers for the end stations through one optical splitter, or a tree and branch topology can be used with a number of splitters. Moreover, this introduces reflection issues that have to be taken into account. Thus, rather than restricting the topologies, a set of optical parameters are used to specify the cable plant.

Second, the optics are 'bursty'. Unlike a point-to-point link, where the optics come up the first time and are, essentially, on thereafter forever, in a point-to-multipoint system, only the CO optic can continuously transmit in the downstream direction. In the upstream direction, the various CPEs have to share the bandwidth via the protocol detailed in the next overview section. This requires that the CPE transmitters in the upstream are capable of switching on and stabilizing, as well as switching off, in a controlled and rapid fashion.

Moreover, the receiver in the CO optic has to be able to adjust rapidly to what could be a widely varying power level in the upstream direction. For instance, a CPE could be at a close distance to the CO with a transmitter that is on the high side of the power range, while the next CPE could be at a far distance to the CO with a transmitter that is on the low side of the power range, thus creating a significant shift in the power level. Again, the receiver needs to adjust in a rapid and controlled fashion.

12.6.8 Optical reach

With the exception of 1000BASE-LX10 operating over MMF, all the optical devices are specified to go at least 10 km. A long distance is expected in an access environment, nonetheless, the astute reader may ask why 10 km, and why are there two distance solutions in EPON, 10 km and 20 km?

The distances were based on data presented by a number of providers and vendors. Outside of North America, 10 km seemed to satisfy a great proportion of needs. Within North America, it was satisfactory for business loops. 20 km, on the other hand, was introduced for residential loops in North America. Having two separate options for EPONs allows for the longer reach solution without having to burden the shorter reach solution with the additional cost on the optics.

12.6.9 Extended temperature

Traditionally, the IEEE had not specified a temperature range for its optics which worked well for a mainly enterprise audience. In the access space, however, the geographic regions, outside plant, business and residential constraints and variations, are endless. For instance, some expressed interest in having the CPE device sit outside the home, while others wanted it inside. Moreover, some had interest in geographic regions, others in cold, and others in both.

The optics themselves were specified over a fairly large temperature range of −40°C to +85°C. Nevertheless, the geographic issues and other constraints would have a cost implication on testing. To accommodate all combinations, a low temperature range, high temperature range, the entire range or no range is permitted, with the latter being no different to the previously unspecified enterprise optics.

12.7 Overview of EPONs: the logic behind P2MP

12.7.1 High-level architecture

As discussed in the P2MP section of the optics overview, EFM introduces a family of solutions that allow sharing of the optics and, to some extent, the fiber, with multiple end stations. In addition to the optical requirements discussed above, a protocol is required.

Put simply, the protocol addresses the issue of having an unknown number (initially) of end stations on a shared fiber plant trying to communicate upstream with one node. An additional requirement on the protocol was to utilize the existing MAC defined in 802.3 with minimum augmentation.

The protocol is known as MPCP (Multipoint Control Protocol) and utilizes the existing feature of using special MAC control opcodes in the MAC control sublayer. A widely used opcode is for PAUSE, or flow control, and is defined in Annex 31A. Figure 12.2 shows the protocol positioning in relation to the MAC and PHY.

12.7.2 A note on terminology

The access literature uses a number of terms to identify the same thing. For instance, head end and CO are used to identify the equipment at the provider's side, while end station and CPE may be used to identify the equipment at the user side.

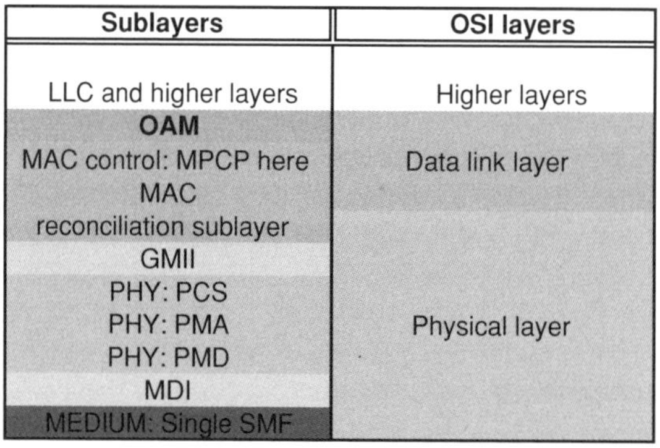

Sublayers	OSI layers
LLC and higher layers	Higher layers
OAM	
MAC control: MPCP here	Data link layer
MAC	
reconciliation sublayer	
GMII	
PHY: PCS	
PHY: PMA	Physical layer
PHY: PMD	
MDI	
MEDIUM: Single SMF	

Figure 12.2 The position of MPCP within the protocol stack

When discussing EPONs, another set of terms is used. Specifically, ONU (optical network unit), which indicates the user side, and OLT (optical line terminal), which indicates the provider side. As with point-to-point devices, the 'downstream' direction is from the OLT to the ONU. The 'upstream' direction is from the ONU to the OLT.

There is never more than one OLT associated with a set of ONUs. In a typical EPON, there will be multiple ONUs.

12.7.3 MPCP: a more detailed discussion

Fundamentally, the protocol has to solve two problems: periodically discovering how many ONUs are on the network, and assigning unique times for each ONU to transmit upstream. The latter also involves synchronizing all the ONUs' clocks, while the first incorporates the distance that each ONU is out.

It is noteworthy to mention that the MPCP provides maximum flexibility to the higher layers in the frequency of discovery and the bandwidth allocation scheme, i.e. it does not constrain either but simply provides the mechanism to control both.

To achieve this, five special frames are defined by the MPCP. They are assigned unique opcodes in Annex 31A, along with the previously defined PAUSE frame. The MPCP may operate with flow control but care should be taken to ensure that the targeted efficiency is achieved. The five frames are as follows:

- *GATE*. This frame is sent by the OLT during normal operation and allows the recipient ONU to transmit frames for a period of time specified in the frame.
- *REPORT*. This frame is sent by the ONU during normal operation and it notifies the OLT of pending transmission requests. Essentially, this is a state dump by the ONU to the OLT.
- *REGISTER_REQ*. This frame is sent by the ONU during the discovery process to initiate the discovery.
- *REGISTER*. This frame is sent by the OLT during the discovery process to inform the ONU that it has been recognized on the network.
- *REGISTER_ACK*. This frame is sent by the ONU during the discovery process to complete the handshake by affirming the receipt of the REGISTER frame from the OLT.

12.7.4 The functional responsibilities of the ONU and OLT

To get a clearer understanding of how the EPON protocol functions, below is a breakdown of functional responsibilities of both ONU and OLT:

1. ONU
   - Operation: synchronizes to the OLT via timestamps on downstream control frames sent by the OLT.
   - Operation: waits for its grants in order to transmit data.
   - Operation: may request additional bandwidth via report frames.
   - Discovery: ONU waits for discovery gate before transmitting a request onto the network.
   - Discovery: ONU acknowledges registration.

2. OLT
- Operation: generates timestamped messages to be used as global time reference.
- Operation: schedules ONU transmission and allocates bandwidth via grants.
- Discovery: generates discovery windows for new ONUs.
- Discovery: performs ranging operation to figure out the distance to the new ONU.
- Discovery: registers the new ONUs.

12.8 Summary

EFM introduces a number of new technologies to Ethernet to address the growing access space. EFM introduces a couple of copper solutions and a set of optical solutions, as well as management and environmental capabilities.

This chapter has addressed each of these solutions individually. Nevertheless, it is important to note that as the access space evolves, EFM provides flexibility beyond the variety of solutions introduced, by allowing for hybrid networks that mix the various copper and optical solutions, but are unified by the Ethernet protocol.

13

DOCSIS as a Foundation for Residential and Commercial Community Networking over Hybrid Fiber Coax

Steven Fulton[1], Chaitanya Godsay[1,2] and Radim Bartoš[2]
[1]*University of New Hampshire, InterOperability Laboratory, Durham, USA*
[2]*Department of Computer Science, University of New Hampshire, Durham, USA*

13.1 Introduction

Residential cable services began in 1948, as an effort to provide television signals to remote locations suffering from poor over-air reception. Early efforts were simply a distribution of radio frequency (RF) signals from transmission stations to the consumer using cables, rather than antennas. Multiple users could connect an antenna outlet from their television set to the common antenna cable running throughout the neighborhood. That 'common antenna' was the genesis of the community antenna television (CATV) system. Cable companies, more formally known as multiple systems operators (MSOs), have grown substantially over the years, both in number of households connected and number of available services. MSOs now provide Internet access to 13 000 000+ homes in the United States (as of June 2003) [1]. That number represents approximately 18 % of the 73 million households receiving a cable television service in the United States.

The principle behind cable distribution systems is straightforward: distribute RF signals from a central location, called the 'head end', to the end subscriber nodes [2]. Figure 13.1 gives the basic layout with coaxial cable connecting the head end to numerous taps that provide CATV connectivity to the subscribers' television sets.

CATV system limitations fall into two general categories: distribution distance and frequency range.

Broadband Services: Business Models and Technologies for Community Networks. Edited by I. Chlamtac,
A. Gumaste and C. Szabó © 2005 John Wiley & Sons, Ltd. ISBN 0-470-02248-5.

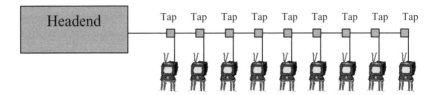

Figure 13.1 Basic CATV distribution

13.1.1 Distribution distance

Signal propagation distance from the head end will vary, depending upon the signal strength and transmission medium; stronger signals travel farther on a transmission line than weak signals. High-quality coaxial cable maintains signal integrity over a longer distance than low-quality coaxial cable. Also, amplifiers can boost weak signal strength at the edges of a network to extend the cable's reach, albeit with slightly less signal quality due to amplifier-introduced signal imperfections.

Fiber optic cable provides lower signal loss than copper cable, permitting even further system distribution without amplifiers. The current cable distribution systems include a combination of coaxial cable with fiber optic cable, and are commonly called hybrid fiber coax (HFC) networks.

Modern HFC cable systems can extend as far as 50 miles from a head end and carry frequencies as high as 860 MHz. Newer cable-based services, such as Internet access and digital telephone, provide interactive connectivity to the end user, so the cable system needs to be bidirectional; able to carry (and amplify) RF signals both to the end user from the head end, and from the end user to the head end.

13.1.2 Frequency range

A single cable channel traditionally requires 6 MHz of bandwidth, so system owners are able to provide more channels when their cable system expands to carry additional bandwidth. Table 13.1 shows some of the frequency assignments for common US television channels.

High-frequency analog signals experience more attenuation in the copper transmission lines than lower frequency signals, so cable systems continually seek to deploy newer models of transmission lines with improved attenuation characteristics for higher frequencies. The resulting cable and amplifier combination, also referred to as the cable 'plant', can support additional channels. Figure 13.2 shows a simplified HFC cable plant.

In 1997, CableLabs, the Colorado based technology arm of multiple MSOs, published a technical standard defining a transmission mechanism to provide data network connectivity through an HFC cable system. This standard established a common technical reference for all vendors developing devices for Internet access over CATV.

The Data-Over-Cable Service Interface Specification (DOCSIS) [3] from CableLabs is one of many last mile technologies intended to provide Internet access and multimedia services to residential users. DOCSIS transports data over a 6 MHz RF channel on the HFC network linking cable modems (CMs) to the cable modem termination system (CMTS). The advent

Table 13.1 Some RF television channel allocations

Channel	Frequency (MHz)
2	54–60
3	60–66
4	66–72
5	72–82
6	82–88
7	174–180
8	180–186
9	186–192
10	192–198
11	198–204
12	204–210
13	210–216
14	470–476
15	476–782

of DOCSIS technology, which uses a shared HFC network medium to provide the 'last mile' of WAN access to commercial businesses and residential homes, requires a new paradigm for WAN interoperability and performance testing. WAN technologies providing last mile connectivity have traditionally used circuit provisioning on a central office (CO) switching device to provide static bandwidth to a circuit termination device at a customer site. The method for performance testing leased line connections has traditionally included an extended bit error rate test (BERT), since the circuit was dedicated to the particular user.

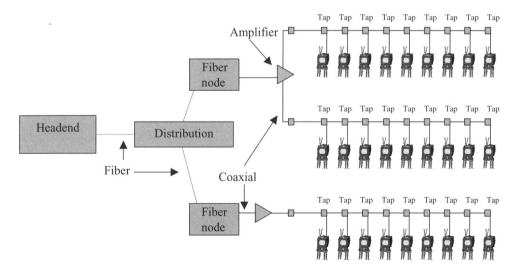

Figure 13.2 HFC distribution

Table 13.2   Internet connection speed growth rates (US)

Speed	May 02	May 03	% Change
Broadband total	26 113 000	38 957 000	49.2 %
Narrowband total	79 444 000	69 647 000	−12.3 %
Modem 14.4K	3 966 000	3 454 000	−12.9 %
Modem 28.8/33.6K	12 014 000	10 118 000	−15.8 %
Modem 56K	63 465 000	56 075 000	−11.6 %

Source: Reproduced from Nielsen//NetRatings, May 2003.

13.1.3 Broadband deployment growth

Broadband is a generic term generally used to describe data circuits with bandwidth capacity in excess of 128 kbps, while narrowband describes 128 kbps and slower circuits. Increasing bandwidth demand is driving a steady migration from narrowband dial-up circuits toward broadband circuits that are continually connected. 63.2 % of the broadband connections in the United States are provided through MSOs offering a DOCSIS service [1].

The results in Table 13.2 show the total broadband and narrowband connection growth in the United States between May 2002 and May 2003. For the 12 844 000 new broadband connections deployed from May 2002 to May 2003, 9 797 000 were conversions from narrowband dial-up services. Only 3 047 000 connections were either completely new broadband connections, or conversions of other, non-dial services.

Even though broadband growth is 49.2 % for the year ending May 2003, less than 25 % of the growth comes from new installations where there is no prior narrowband access. Data extrapolations of Table 13.2, using the 49 % broadband annual growth trend and the 12 % narrowband annual conversion trend, show that to maintain the current 49.2 % broadband growth rate, the broadband market must attract over 10 000 000 new connections in 2004 that are not merely narrowband conversions, see Table 13.3.

13.1.4 Broadband deployment future

The results in Table 13.3 indicate that the broadband industry must increase the rate of narrowband conversion and find new broadband sales markets if current growth rates are achievable beyond 2003. Some of the competing broadband technologies, namely DOCSIS and DSL, are already working to maintain growth rates through cannibalization of the competitive technology's installed user base [4]. DSL providers have traditionally sought to serve the business markets. DOCSIS providers have traditionally sought to serve the residential markets already receiving cable video services. The DOCSIS industry is now targeting small business data services as new opportunities [5]. The result will be business-related narrowband conversions, alternatives for existing xDSL customers, conversions of leased line service, and further efforts to deploy corporate VPN services across the DOCSIS environment.

Of particular interest for DOCSIS service providers will be the circuit turn up process for any business customers migrating from leased line services to DOCSIS services. Leased line

Table 13.3 Extrapolation of broadband growth components (US)

	Growth rate	May 02	May 03	May 04	May 05	May 06	May 07
Broadband connections	49.2 %	26 113 000	38 957 000	58 123 844	86 720 775	129 387 397	193 045 996
Narrowband connections	−12.3 %	79 444 000	69 647 000	61 080 419	53 567 527	46 978 722	41 200 339
Conversions (narrowband to broadband)			9 797 000	8 566 581	7 512 892	6 588 806	5 778 383
New broadband connection or conversion of 'other' existing service			3 047 000	10 600 263	21 084 040	36 077 816	57 880 216
How much of the 49.2 % broadband growth rate must come from new non-narrowband connections			24 %	55 %	74 %	85 %	91 %

services are provided with steady-state bandwidth capacity, verified during installation using an unframed BERT. Since DOCSIS provides shared access to the HFC for multiple users, it is impossible to conduct an unframed BERT on an active channel. A business user will need to be certain that critical applications are using a guaranteed bandwidth reservation system through DOCSIS Quality of Service (QoS) mechanisms, as described in Section 13.2.3.2, and that a framed BERT is performed to verify the expected bandwidth capacity.

13.2 Provisioning process for DOCSIS connections

Cable operators, in the early 1990s, envisioned the growth of cable networks and were driven to explore possibilities for transmitting data from the residential user to the service provider. By providing this capability, packet-based services, such as high-speed Internet access, cheaper telephone connections and videoconferencing could be deployed easily. This led to the formation of many research groups, and Multimedia Cable Network System (MCNS), a collaboration of cable companies, was the first to come up with a specification. MCNS released the set of standards known as DOCSIS 1.0 (Data-Over-Cable Service Interface Specification) in March 1997. CableLabs, a nonprofit research and development consortium, worked in collaboration with MCNS and is now responsible for developing new specifications and product certification.

The DOCSIS specification [3] describes a DOCSIS network as a tree-based network with the cable modem termination system (CMTS) as the root of the tree and the cable modems (CMs) as the leaves of the tree. The CMTS is at the service provider facility and the CMs are at the residential users' homes. The transmission of data from the CMTS to a CM, termed downstream, is a point-to-multipoint broadcast, whereas the transmission from the CM to a CMTS, termed upstream, is controlled by the CMTS and is multipoint-to-point TDMA (time division multiple access). DOCSIS defines an asymmetric network in terms of upstream and downstream data rate, with downstream rates (up to 30 Mbps) being substantially larger than the upstream rates (up to 10.24 Mbps). Data transmission in DOCSIS is full duplex. A simple DOCSIS network is shown in Figure 13.3. The residential user has the customer premises equipment (CPE), such as computer, telephone, etc., connected to the CM. Upstream data goes from the CM to the CMTS, and is then forwarded appropriately to the outside network. Similarly, downstream data passes through the CMTS to the CM, and is forwarded to the CPE. Typically, there are 1500 to 2000 CMs connected to a CMTS, with distances between the CMTS and CMs going up to 50 miles. The DOCSIS network is also known as the radio frequency (RF) network and the hybrid fiber coax (HFC) network.

After the DOCSIS 1.0 specification, CableLabs released two more specifications, known as DOCSIS 1.1 and DOCSIS 2.0. DOCSIS 1.1, released in 1999, enhanced security and added Quality of Service (QoS) to support real-time applications, such as telephony and videoconferencing. DOCSIS 2.0, released in 2001, provided a significant upstream capacity upgrade, necessary due to the increased number of users on the cable network, and enabled new high-speed applications, such as peer-to-peer file sharing and online gaming. This upgrade also aimed to provide better noise immunity and support even more users than before. So the dream of having one wire running into a residence that provides high-quality features such as digital audio and video, provides real-time and interactive support for telephony, videoconferencing, online gaming, and high-speed Internet connectivity for browsing and peer-to-peer file sharing

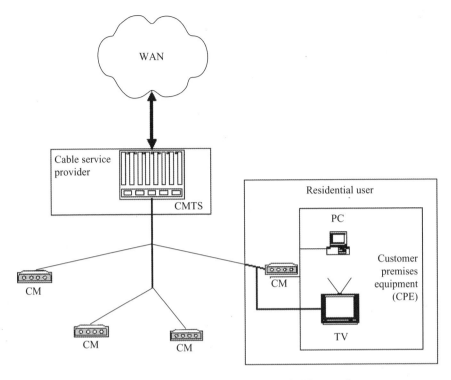

Figure 13.3   Architecture of a DOCSIS RF network

can be met through DOCSIS. These are some of the applications that have benefited because of the high speed provided by DOCSIS. DOCSIS has concentrated on providing the base for supporting various kinds of QoS. Thus, more and more innovative applications will be developed to harness the full potential of DOCSIS.

The next section discusses the DOCSIS protocol for cable modem provisioning, followed by a discussion on upstream and downstream data transmission.

13.2.1 DOCSIS provisioning basics

Before a cable modem (CM) can start transmitting data upstream towards the head end device in the multiple service operator (MSO) central office, the CM must complete a nine-step process involving signal exchanges with the head end cable modem termination system (CMTS) device. Once the CM finishes the start-up interaction with the CMTS, it has joined the network and can request to transmit data upstream to the CMTS. This process, which the CM completes in order to join the network, is called ranging and registration. A variety of CM operational parameters are configured during the ranging and registration process to allow for CM device management, and to specify bandwidth transmission capabilities for the CM. The next section describes the ranging and registration process and also introduces the important message flows in a DOCSIS network which are required to understand the DOCSIS protocol operation.

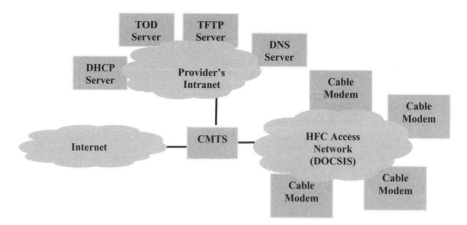

Figure 13.4   System devices for DOCSIS connection provisioning

For proper DOCSIS provisioning, a number of ancillary services, such as the Dynamic Host Configuration Protocol (DHCP) to provide IP addresses, Trivial File Transfer Protocol (TFTP) to provide operational configuration to the CM, time of day (TOD) servers to provide accurate time, and domain name server/service (DNS) for name resolution, must be present, in addition to the CM and CMTS, as shown in Figure 13.4.

13.2.2 Ranging and registration

The following steps outline the process for bringing a CM online in a DOCSIS system.

1. CM scans the available downstream frequencies coming from the CMTS to find a data channel and synchronize the CM receiver with the SYNC pulses present on the data channel.
2. CM listens to the downstream frequency until it receives an upstream channel descriptor (UCD) packet, which will specify upstream transmission parameters necessary to communicate back to the CMTS. (Parameters include a specific upstream frequency, modulation type, symbol rate, channel ID, preamble pattern and burst descriptors).
3. Ranging with the CMTS to negotiate the appropriate upstream power levels and latency offsets for transmission delay.
4. Identify a device class identifier that tells if the cable modem is embedded in a PC.
5. Establish IP connectivity for the CM by obtaining an IP address from the DOCSIS DHCP server.
6. Establish time of day with the time of day server specified during DHCP configuration.
7. Obtain operational CM parameters from a configuration file stored on the TFTP server specified during DHCP configuration.
8. Register with the CMTS so that it can assign an identifier to the downstream data flow to the CM and the upstream data flow from the CM.
9. Initialize baseline privacy interface (BPI), which involves negotiating encryption keys used to encrypt data between the CM and CMTS.

It is important to understand the steps for the range and registration process, in order to troubleshoot basic DOCSIS device connectivity problems. It is equally important to understand the detail associated with each step in the outline, because the root of device interoperability problems arises within the procedural details, not at the outline level. A slightly more detailed discussion of the range and registration process follows.

Step 1. When a CM has to join a DOCSIS network after power on, it first searches for a valid downstream signal in the frequency range of 91–857 MHz. A downstream signal is valid when the CM has recognized and synchronized with the PHY layer (QAM symbol timing, FEC framing), and with the MAC layer (MPEG packetization and SYNC messages). A CM is said to be 'MAC synchronized' when it receives at least two SYNC messages within the clock tolerance limits.

Step 2 and 3. After MAC synchronization, the CM must negotiate the appropriate transmit power level with the CMTS and determine the latency for transmission between the two devices. The CM now learns about the upstream channel characteristics using an upstream channel descriptor (UCD) sent by the CMTS at regular intervals. Once it has learned about the upstream channel characteristics, it knows how to transmit data upstream, i.e., modulation formats, modulation rates, etc., however, it does not know when to transmit. The CMTS periodically sends information regarding when the CM is allowed to transmit in a MAC message termed the MAP.

The MAP is a MAC message that indicates when a CM can send data. A MAP message has grants for initial maintenance (IM), station maintenance (SM), request (REQ), request/data (REQ/DATA), short data grant and long data grant. A CM uses the initial maintenance region to join the network. The initial maintenance (IM), request (REQ) and request/data (REQ/DATA) are broadcast regions, meaning that any CM can try to transmit information upstream in this region and the information is subject to collisions. The station maintenance (SM), short data grant and long data grant are unicast opportunities provided to the CM by the CMTS and are not subject to collisions.

The CM then transmits a request to join the network (RNG-REQ) in the initial maintenance region. There is a possibility of collision in this region if multiple CMs want to join the network. To resolve collisions, an exponential back-off scheme is used. On successful transmission of the RNG-REQ, the CM gets a unique identifier from the CMTS, known as the service identifier (SID). The SID is sent in a message called the range response (RNG-RSP). The CM then uses this SID for all future communication with the CMTS. RNG-RSP also gives the physical layer adjustment information, such as timing adjust, power adjust, frequency adjust, etc., to the CM. The CM also gets unicast station maintenance opportunities in the MAP after the first RNG-RSP. The CM then has to keep on fine-tuning the physical layer adjustments until the CMTS is satisfied and sends the ranging complete status in the RNG-RSP. This process of fine-tuning is known as station maintenance, and goes on at regular intervals due to the tendency of the physical devices to go out of synchronization for various reasons. The CMTS sends unicast transmit opportunities to the CM to transmit RNG-RSP messages upstream, when requested by the CMTS, in the station maintenance region of the MAP.

Step 4. This step is used to detect CMs that are embedded in other devices, such as a CM that resides on a peripheral card within a computer chassis. Although there are no commercial products fitting this description, the DOCSIS specification allows for a computer-controlled cable modem (CCCM) that uses computer processor and memory in an effort to tightly couple the CM and computer resources.

Steps 5, 6, 7 and 8. The CM requests an IP address from the DHCP server, establishes time of day with the TOD server and downloads the configuration file from the TFTP server. The configuration file specifies all the functionality the CM can support. After getting the configuration file, the CM must send a registration request (REG-REQ) to the CMTS. The REG-REQ message enlists all the functionality the CM wishes to support, as specified in the configuration file. The CMTS responds using a registration response (REG-RSP) message, indicating to the CM all the functionality the CM is allowed to support. The CMTS makes the final decision regarding what functionality will be operational in the CM.

Step 9. If the CM configuration file specifies that the modem is to use the baseline privacy interface (BPI) protocol to encrypt communication between the CM and CMTS, then the last step to bring the modem online is to go through a secure key exchange process to enable BPI.

At this point the CM has successfully ranged and registered with the CMTS.

13.2.3 Upstream data transmission

Once the CM has registered with the CMTS, it is ready to send data in the upstream direction. It knows how to send data upstream, as specified by the UCD sent by the CMTS at regular intervals. It is synchronized with the CMTS in time using the SYNC messages and ranging. However, it does not know when to send data.

Transmitting data upstream is a three-step process, as shown in Figure 13.5. When the customer premises equipment (CPE) attached to the CM needs to sends data onto the network through the CM, the CM looks in the most recent MAP for the REQ or REQ/DATA region. It then generates a data grant request message to the CMTS indicating the grant size and tries to transmit the request in the time specified for the REQ or REQ/DATA region. As mentioned before, these regions are subject to collisions since many CMs may be trying to send data grant request messages. If the data grant request message reaches the CMTS, then the CMTS either sends a long or short data grant to the CM in the following MAP. Whether a long or a short data grant is sent depends on how much the CM has requested in the data grant request message. The CMTS will send a data grant pending message in the MAP when the CMTS has received the data grant request from the CM but cannot allocate data grant to it. So the CM detects a collision if it does not see a short data grant, a long data grant or a data grant pending in the next MAP, and increases the window size of the exponential back-off algorithm to resolve the collision. The CM will then defer a certain number of request opportunities before requesting again.

If the CM gets a short/long data grant successfully from the CMTS, then it extracts the time to send the data from the MAP. Finally, at the time to send data, the CM transmits the data to the CMTS. This is termed the request-data grant-send (RDS) cycle. So the CM has to go through one or many RDS cycles (in case of collisions) to transmit data upstream.

A typical DOCSIS network may have up to 1500–2000 CMs on a CMTS with distances ranging up to 50 miles, and with the CMs being quite close to each other distribution-wise. All these factors lead to a high probability of collisions in the REQ or REQ/DATA region. The

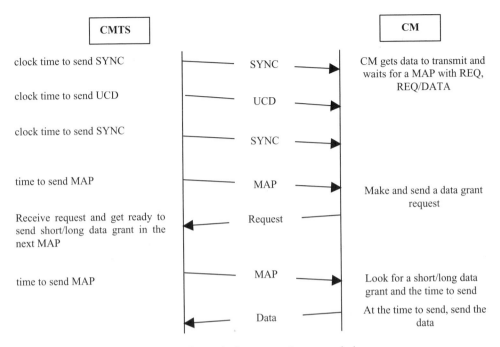

Figure 13.5   Typical upstream data transmission

time required, or the time delay, in the RDS cycle is thus the real bottleneck in the upstream throughput performance.

Thus, more delay in the RDS cycle means more waiting for new packets and more packets being dropped (as the CM cannot request another data grant if a previous request is pending), thus leading to smaller throughput and higher latency.

To improve the throughput and latency of the upstream there are several performance enhancers, such as concatenation, piggybacking and fragmentation, and a set of QoS services, such as an unsolicited grant service (UGS), an unsolicited grant service with activity detection (UGS-AD), real-time polling (rtPS), and non real-time polling (nrtPS), that aim to control the delay in the RDS cycle. These parameters are set by using the configuration file that the CM downloads during registration.

13.2.3.1 Performance enhancers

Concatenation
If the CM has several packets to send it can save on time for multiple RDS cycles per packet and also on per packet overhead by concatenating and sending many packets together. The CM must, however, know the final concatenated packet size and request for a data grant accordingly.

Piggybacking
The CM, while sending the data upstream at the time allocated by the CMTS, can put an extended header on the data packet to request additional data grants. The CM totally bypasses

the collision phase in the RDS cycle this way and can keep on requesting additional data grants as long as it has data to send.

Fragmentation

There can be times when the CMTS can give a smaller data grant than that requested by the CM. In this case, the CM can either not use the data grant and request again, or it can fragment the packet and send as much as possible in that grant. The CMTS will send a smaller grant when it is overloaded with data grant requests from many CMs.

13.2.3.2 QoS services

The DOCSIS 1.1 and 2.0 specifications [3, 6] support a series of scheduling service guarantees, as required by the applications.

Unsolicited grant service (UGS)

This service is supported for real-time traffic, characterized by fixed size packets at fixed intervals, such as VoIP telephone calls. It directly gives data grants to the CM at regular intervals, thus completely bypassing the RDS cycle.

Real-time polling service (rtPS)

This service is supported for real-time traffic, characterized by variable size packets at fixed intervals, such as MPEG video. In this case, the CMTS sends unicast REQ opportunities to the CM, thus bypassing the collision phase of the RDS cycle. The CM then requests the data grant size it wants and gets to transmit the data within the specified interval.

Unsolicited grant service with activity detection (UGS-AD)

This service is supported for real-time traffic, characterized by an inactive state for a substantial period of time followed by a period of fixed size, a fixed interval active state. A VoIP call with silence suppression is a good example that uses this service. This can be imagined as a combination of RTP and UGS, where the CMTS sends unicast poll/request opportunities to the CM in the inactive state and sends fixed size data grants at regular intervals in the active state.

Non real-time polling service (nrtPS)

This service is supported for non real-time traffic, characterized by variable size packets at regular intervals, such as high-bandwidth FTP. The CMTS just provides unicast request opportunities to the CM at regular intervals. This ensures that the CM gets at least some request opportunities in cases of network congestion.

Best effort (BE)

This is not a guaranteed service, the CM has to go through the RDS cycle and the CMTS does its best to provide as many data grants as it can. Internet browsing applications are generally put on a best effort scheduling service. This is the default service when no other scheduling service is specified.

The most recent DOCSIS 2.0 specification has provided a significant bandwidth upgrade and increased the upstream capacity to 30 Mbps.

13.2.4 Downstream data transmission

The transmission of data from the CMTS to CM is called downstream data transmission. As video signals are sent in the downstream direction as MPEG packets, data in the downstream direction is also encapsulated as MPEG packets to multiplex video and data effectively. This also facilitates the same receiving hardware for video and data. The traffic in the downstream direction is all 188 byte MPEG traffic, and DOCSIS packets can be identified by packets having type 0x1FFE. The CMTS is responsible for multiplexing video and data. Thus, packets coming in from the outside network are split if required, converted into MPEG packets and transmitted downstream.

As upstream transmission is controlled by the CMTS, management packets (SYNC, MAPs, UCDs, etc.) are also sent downstream, in addition to data. DOCSIS 1.0 and 1.1 provide downstream data rates from 10 Mbps to 30 Mbps [7, 8]. However, DOCSIS 2.0 delivered a significant downstream bandwidth upgrade ranging from 40 Mbps to 110 Mbps.

13.3 Summary

The broadband market is growing, but is largely composed of narrowband conversions. Current growth rates cannot be sustained unless broadband finds new markets beyond narrowband conversions.

DOCSIS provisioning requires significant back office support systems, so DOCSIS testing must be cognizant of dependencies on these back office system implementations. DOCSIS testing needs to continually look beyond the DOCSIS protocol itself, in order to address other embedded technologies residing in DOCSIS devices, such as spanning trees, Ethernet and 802.11 wireless technologies. DOCSIS bandwidth management was originally on a best effort basis, but business critical applications require service level agreements and are not amenable to a best effort network reservation mechanism.

DOCSIS 1.1 introduced QoS mechanisms that can provide guaranteed bandwidth reservation, but the QoS mechanisms need MSO operational implementation in order to address business network needs. DOCSIS 2.0 offers additional modulation choices to increase transmission capacity, as well as to improve signal integrity during degraded transmission conditions.

The DOCSIS protocol provides a viable mechanism for last mile access to both residential communities and small businesses, provided that the local provider properly implements the advanced QoS features, along with the newer modulation features provided in DOCSIS 2.0.

References

1. Harris, M. (2003), *U.S. Broadband Subscriber Count Tops 20 Million*, in Cable Datacom News (Online document), available at: http://www.cabledatacomnews.com/sep03/sep03-1.html.
2. Adams, M. (2000), *Open Cable Architecture*, Cisco Press, Indianapolis.
3. DOCSIS (2003), *Data-Over-Cable Service Interface Specifications: Cable Modem Radio Frequency Interface Specification SP-RFIv1.1-I10-030730*, http://www.cablemodem.com.
4. Burstein, D. (2003), *The Inside Source*, in DSL Prime News (Online document), available at: http://www.isp-planet.com/cplanet/tech/2003/prime_letter_030109.html.
5. *Small Business Solutions*, (Online document), available at: http://www.cable-modem.net/pi/ business_solutions.html.

6. DOCSIS (2002), *Data-Over-Cable Service Interface Specifications: Cable Modem Radio Frequency Interface Specification SP-RFIv2.0-I03-021218*, http://www.cablemodem.com.
7. Godsay, C. K. (2003), *Upstream Performance Whitepaper*, InterOperability Laboratory, University of New Hampshire, Durham, New Hampshire.
8. Bartoš, R., Godsay, C. K. and Fulton, S. (2004), Experimental Evaluation of DOCSIS 1.1 Upstream Performance, *in Proceedings of the IASTED International Conference on Parallel and Distributed Computing and Networks (PDCN)*, Innsbruck, Austria.

14

Broadband Wireless Access Networks: a Roadmap on Emerging Trends and Standards

Enzo Baccarelli[1], Mauro Biagi[1], Raffaele Bruno[2], Marco Conti[2,3] and Enrico Gregori[2,3]

[1]INFO-COM Department, University of Rome 'La Sapienza', Rome, Italy,
[2]Istituto IIT, National Research Council (CNR), Pisa, Italy,
[3]CreateNet Research Consortium, USA

14.1 Introduction

Broadband wireless communication is expected to expand social and economic benefits arising from entertainment, e-commerce, teleworking, e-learning and even e-government. It may also be the desired bridge toward the ubiquitous and pervasive utilization of the Internet platform. However, broadband communications requires higher capacity for the access networks linking end users to the backbone infrastructures. Although actual access connections are often adequate for large organizations (such as business corporations and universities), nevertheless broadband (possibly reconfigurable) access networks for residential users are also needed for achieving the full social and economic promise of the 'broadband paradigm'.

Under this perspective, the aim of this chapter is twofold. First, it provides a snapshot of current technologies, standards and regulation policies involved in the Broadband Wireless Access (BWA) paradigm. Secondly, it explores (and somewhat anticipates) some emerging BWA technological trends, future BWA potential benefits, projected demands, expected progress and challenges for the multiple (often contrasting) technological options envisioned for providing BWA to the end users. Future trends of BWA's alphabet soup – such as wireless personal area networks (WPANs), wireless local area networks (WLANs), wireless metropolitan area networks (WMANs) and satellite-based systems – are presented and compared.

Broadband Services: Business Models and Technologies for Community Networks. Edited by I. Chlamtac,
A. Gumaste and C. Szabó © 2005 John Wiley & Sons, Ltd. ISBN 0-470-02248-5.

Key questions in relation to this chapter include:

- What is the BWA and do people need it?
- How much demand is there for BWA?
- What are the current technological paradigms and standards for BWA?
- What are the emerging technologies able to support incoming BWA networks and applications?
- What are the expected Quality of Service (QoS) parameters supportable by emerging next generation BWA platforms?

It is likely that the multifaceted and dynamic picture emerging from this contribution will also supply a useful roadmap to regulators and policy makers.

14.2 What is Broadband Wireless Access?

The term Broadband Wireless Access (BWA) has become commonplace for denoting future radio digital communications. At face value, the term 'broadband' refers to the *large* bandwidth that a high-speed connection should guarantee to the end users. But is there a well-defined threshold above which a service is broadband? And is the bandwidth the only requirement demanded to classify a service as 'broadband'?

In recent years, various working groups and standardization bodies have attempted to develop appropriate definitions of BWA, and these definitions have changed over time. From the end user perspective, 'broadband access' is perceived as more than a large bandwidth. Multiple heterogeneous working groups would benefit from a *workable* definition of what constitutes BWA. These groups include consumers, service providers, application and content developers, policy makers and standardization bodies. Thus, framed in this way, defining the term 'Broadband Wireless Access' is not limited to specifying a minimum access rate, but also involves identifying the 'killer' applications potentially most appealing for the end users and evaluating the technical and social benefits that different segments of the public anticipate from wireless access to broadband services.

14.2.1 QoS requirements for BWA

Large access rates (or high access speed) are only one of the technical features demanded by BWA. Along with access speed, always-on connectivity, latency/jitter, bandwidth symmetry and addressability are also important performance components involved in the definition of BWA. The related technical challenges are briefly described in the following.

- *High-speed access.* Although the access speed of a service (e.g., the data rate at which an end user can transfer data to and/or from his electronic equipment) depends on multiple networking factors, it is recognized that a threshold ranging from 2 to 5 Mbps for the end user access rate would seem adequate for most envisioned future 'broadband' applications.
- *Latency and jitter.* While the access rate may be the most significant of many broadband applications, two additional parameters, namely, latency and jitter, may become crucial for real-time broadband services, such as audio/video streaming and videoconferences. Latency

affects applications requiring interaction, such as human-to-human conversations, games and so on. Jitter measures the latency fluctuations, and several broadband services (such as, for instance, video streaming) are very jitter sensitive.

- *Symmetry between uplink and downlink access rates.* Currently, most narrowband telecom services do not guarantee the same uplink and downlink access rates. However, although the term broadband does not necessarily imply symmetric uplink and downlink access rates, nevertheless several broadband services (digital telephony and videoconferencing being the most mature) are emerging that place increasing demand on the upstream access rates.
- *Always-on connectivity.* Today (e.g., before the pervasive advent of broadband services), residential Internet users are typically confined to using a dial-up connection to the Internet that forces the users to suffer long access delays. By eliminating the need to place a dial-up connection, future broadband services will greatly reduce the delay. It is anticipated (and, to a certain degree, already experienced) that a reduced access delay will change the way that users perceive and use the Internet and will promote a more pervasive use of the network for short tasks, for example instant messaging.
- *Addressability.* A critical requirement of several emerging broadband applications is that the user's terminal be addressable. This means that someone on the Internet can initiate communications with the user. In addition, addressability also enables some advanced functionality, such as a user being able to run a server which other Internet users can access (a capability demonstrated by Napster and Gnutella). However, addressability is the exception for residential users today, the main technical reason being that the Internet Protocol does not provide dynamically assigned addresses to residential users.

14.2.2 A 'workable' definition of BWA

The above list of technical requirements and challenges suggests the pitfalls of picking an access speed threshold to define BWA. Therefore, the current trend is to define a broadband service as one that meets a specified set of technical requirements, requirements that may change over time. These considerations motivate the following workable definition of BWA.

> *Definition*: BWA should provide sufficient technical performance, and wide enough service penetration meeting the performance level, to promote the development of new applications.

This definition implies that BWA systems are defined both by technical and economic requirements that will allow them to implement a complete chicken-and-egg application cycle.

14.2.3 BWA in personal, local, metropolitan and wide area networks

Depending on their coverage, we may classify future wireless networks into four main segments: personal (PANs), local (LANs), metropolitan (MANs) and wide area networks (WANs). Traditionally, broadband services refer to new generations of high-speed data services aimed at residential and small business users. Therefore, the scope of BWA systems seems to be restricted to LANs, MANs and WANs, which implement wireless access services on a metropolitan or nationwide scale. However, we can define a BWA system also as a network solution promoting the development of new applications that fulfill the QoS requirements detailed in Section 14.2.1. From this point of view, among the new applications that rely on the availability of broadband services, we should insert ambient intelligence. Ambient intelligence represents the long-term evolution of the ubiquitous computing concept. In this vision, the world around us (homes, offices, cars and cities) is organized as a pervasive network of intelligent devices that will cooperatively gather, process and transport information, forming a virtual world (cyber-world) filled with avatars. In the cyberworld, users' avatars stay in touch with other synthetic agents in order to get services and perform transactions. Virtual immersive communications will constitute the basis for bridging the real and virtual worlds. In this environment, ultra high-speed multimedia communications, (i.e., broadband communications) are required to support, in the most natural way, the interactions between real users and avatars [1].

In the following we outline the most significant standardization efforts performed to specify wireless technologies enabling broadband services in PAN, LAN, MAN and WAN environments. Specifically, we adopt a classical networking perspective: we begin by considering networks with the smallest scope (WPANs) and we analyze the wireless technologies that allow enlargement of the coverage area, i.e. WLANs, WMANs and finally WWANs.

14.3 Technologies for WPANs: the 802.15 standards

Wireless personal area networks (WPANs) are used to transmit information over relatively short distances (< 10 meters) among small groups of digital devices. The development of standards for short-range wireless networks is currently ongoing in the framework of the IEEE 802.15 Working Group for wireless personal area networks. Specifically, the IEEE 802.15 Working Group aims to develop standards for the wireless networking of portable and mobile computing devices, such as PCs, personal digital assistants (PDAs), peripherals, cell phones, pagers and consumer electronics [2, 3, 4].

Currently, four Task Groups (TGs) are active inside the IEEE 802.15 Working Group:

1. TG1 that, in 2002, approved a wireless personal area network standard based on the Bluetooth v1.1 Specifications [5].
2. TG2 that is developing recommendations to guarantee the coexistence between 802.11 and 802.15 devices.
3. TG3 that has just produced a draft standard for high-rate (20 Mbit/s or greater) WPANs. TG3 also contains an Alternative PHY Task Group (TG3a) that is working to define a higher speed physical layer enhancement to 802.15.3 for applications that involve imaging and multimedia. All the alternative physical layers under consideration employ ultra wide band (UWB) communications.
4. TG4 that is investigating a low-data-rate, low-cost and low-power solution for WPANs operating in an unlicensed frequency band. Potential applications are sensors, interactive toys, smart badges, remote controls and home automation.

Hereafter, only the basic concept of 802.15.3 and 802.15.4 WPANs [4] will be introduced, since TG3 and TG4 standards are currently not yet defined.

14.3.1 IEEE 802.15.1: a Bluetooth-based WPAN

The IEEE 802.15 Working Group for wireless personal area networks approved in 2002 its first WPAN standard [2], which is based on the lower part of the protocol stack designed in the

Bluetooth v1.1 Specification. Bluetooth specifications are released by the Bluetooth Special Interest Group (SIG), which is a standardization forum consisting of leading companies in the telecommunications, computing and networking sectors [6]. The Bluetooth SIG aims to define specifications for low-cost, short-range radio links, with data rate up to 1 Mbps, between mobile PCs, mobile phones and other portable devices.

A *piconet* is the fundamental building block of a Bluetooth network. The piconet contains up to eight active (i.e. participating in data exchange) stations, one station has the role of master, while the other stations are slaves. A Bluetooth piconet has a small coverage area (i.e. an area up to 10 meters around the master) that can be extended by interconnecting independent piconets that have overlapping coverage areas. The Bluetooth specification defines a method for the piconets' interconnection: the *scatternet*. A scatternet can be dynamically constructed, in an ad hoc fashion, when some nodes belong at the same time to more than one piconet (inter-piconet units). In general, a scatternet exists when a unit is active in more than one piconet[1] at the same time. Scatternets can be useful in several scenarios. For example, we can have a piconet that contains a laptop and a cellular phone. The cellular phone provides the access to the Internet. A second piconet contains the laptop itself and several PDAs. In this case, a scatternet can be formed with the laptop as the inter-piconet unit. By exploiting the scatternet, the PDAs can exploit the cellular phone services to access the Internet. The current Bluetooth specification only defines the notion of a scatternet but does not define the mechanisms to construct the scatternet.

14.3.2 High-rate WPANs: the IEEE 802.15.3 standard

The aim of Task Group 3 (TG3) is the specification for high-rate WPANs able to meet the requirements of applications performing video and audio distribution (e.g., high-speed digital video transfer from a digital camcorder to a TV screen, home theater, interactive video gaming, etc.), and/or high-speed data transfer (kiosks for still images, MP3 players, printers and scanners, etc.). This task group has already defined the draft version of MAC and 2.4 GHz PHY layers for WPANs with data rates up to 55 Mbps [3]. In detail, with the current PHY, five data rates are possible: 11, 22, 33, 44 and 55 Mbps with a short-range coverage ($\approx$ 10 meters). The 802.15.3 data rate is not enough for supporting highly demanding multimedia applications, such as home theater, interactive applications (e.g., gaming) and high-rate content downloading. For this reason, an additional task group, named IEEE 802.15.3a, was created inside TG3 to investigate a physical layer alternative to the 802.15.3 PHY. The aim is to develop a very high-rate PHY that will replace the IEEE 802.15.3 PHY, while maintaining the same MAC layer. Specifically, TG3a is aimed at defining a physical layer able to support very high data rates on short distances, i.e. 110 Mbps at 10 m, 200 Mbps at 4 m and 480 Mbps at unspecified distances.

Currently, TG3a is comparing and contrasting several alternative proposals for a very high-speed PHY. In the meantime, the FCC in the USA have approved ultra wideband (UWB) devices, and this makes the UWB technology a promising technology for the 802.15.3a PHY. Indeed, many manufacturers have proposed to 802.15.3a a UWB-based physical layer [4, 7]. For this reason, it is worthwhile to briefly sketch the main features of this technology [8].

---

[1] A unit can be master to only one piconet.

UWB is a technique to spread the signal over an exceptionally large bandwidth (well beyond CDMA systems). In the original approach, UWB is based on carrierless impulse radio, and then it uses very short-duration (a few nanoseconds) baseband pulses spanning bandwidths of several GHz. Pulse repetition frequency (PRF) can range from hundreds of thousands to billions of pulses/second. UWB is a form of extremely widespread spectrum where the available energy is spread over a spectrum of several GHz. Hence, in principle, UWB signals should be designed to look like imperceptible random noise to conventional radios. The FCC established that UWB devices could operate with a low power in the unlicensed spectrum from 3.1 to 10.6 GHz. The low emission limit should guarantee that UWB devices do not cause harmful interference to other existing radios, e.g., IEEE 802.11a. To further reduce coexistence problems with existing systems, a new recent approach to UWB is based on a multibanded system [8]. In this case, the UWB band (3.1–10.6 GHz) is partitioned into smaller bands exceeding 500 MHz, and UWB communications occur inside each band. Coexistence with other systems is easily achieved by not using bands where other systems are already present. The multiband approach has additional benefits: it simplifies the technical challenges, and it makes it easier to adapt to the regulatory requirements of different countries. For all these reasons, multiband UWB currently seems to be a viable choice.

14.3.2.1 Current challenges and future directions

Due to both radiated power levels and an extremely wide frequency spectrum, current UWB transceivers suffer from (very) limited coverage ranges, especially in high-performance applications requiring throughputs exceeding 100 Mbps and target error probabilities around 10^{-6} [9]. In principle, substantial coverage gains could be achieved by resorting to multi-antenna physical platforms equipped with suitable space–time codes (STCs) [10]. Although existing UWB modems are indeed single-antenna, transceiver architectures for multi-antenna impulse radio communications are emerging that promise about a factor of 10 increment in the allowable coverage ranges [9]. An additional challenge of UWB systems concerns the interaction between Physical and MAC layers. As stated before, TG3a is working with the constraint of using the 802.15.3 MAC on the top of an 802.15.3a PHY. However, the MAC has several functionalities that are not easily supported with a UWB-based PHY, such as the identification of idle channels (in UWB the energy is not a good measurement for a busy channel), supporting asynchronous broadcast frames in contention access periods, scalability issues (how to extend the size of piconets by interconnecting two or more piconets), etc. These problems and possible solutions are addressed in [7].

14.4 Emerging 4G WLANs

Many visions exist on the meaning of the term 4G wireless systems. On one hand, 4G is used to indicate the air interface evolution that in forthcoming years should allow us to meet higher data rates from 100 Mbps up to 1 Gbps. We expect to achieve this by using technologies such as multi-antenna platforms, improved signal processing techniques, the use of OFDM as a multiple access paradigm, improved signal processing schemes and software-defined radio [11]. On the other hand, 3G evolution is also based on the vision of cooperating wireless networks, with wide area cellular networks, fixed broadband wireless access and wireless local area networking (WLAN) solutions constituting the components of a multistandard heterogeneous network architecture. In this sense, the objective of the wireless networks beyond 3G is to allow users to connect through a multiplicity of devices, anywhere, at a wide range of speeds, rather than simply providing megabits of bandwidth per user. In this scenario, the WLAN technologies play a crucial role because they are the ideal complement to other broadband and cellular wireless technologies for providing cost-effective high-speed nomadic connections in isolated and highly populated areas. Moreover, the WLAN technologies can effectively open up the widespread deployment of wireless Internet applications since they can be used:

- *At the office.* As the wireless extension of the local area network, WLAN is providing mobility for enterprise users.
- *In the home.* An Asynchronous Digital Subscriber Line (ADSL) service can run anywhere in the home, with multiple users connected over a WLAN [12].
- *In the public environment.* WLAN access points are being deployed in so-called 'hotspots', such as cafes, retail shops, convention centers, airports and other areas, where people can benefit by seamless public access to the Internet.

Although the number of public WLANs worldwide is expected to rise to more than 24000 by 2004, there are key business and technical issues that should be solved to drive the adoption of hotspots in public areas. The provisioning and billing models for the hotspots are still immature in terms of how the hotspot operators derive revenue from the public service they offer. Furthermore, the security problem is a significant concern in the deployment of public access services, since the intrinsic accessibility of the wireless technologies could enable eavesdroppers and unrestricted access to the broadband service [13]. However, at least in the first stage of the hotspot deployment, the most crucial concerns for the service providers and the hotspot operators are: (i) how to provide efficient support for services with a wide variety of QoS requirements; and (ii) how to guarantee a true high-speed connection to Internet services as the number of users increases. In the remainder of this section we outline the design principles used by 802.11 technology, which is the most mature technology to implement WLAN solutions, and the current trends to extend it to fulfill the envisioned needs of emerging 4G scenarios.

14.4.1 Technologies for 4G WLANs: the IEEE 802.11 standards

The recent and rapid surge in WLAN usage was fueled by the standardization and proliferation of readily interoperable wireless adapters and hubs. The Institute of Electrical and Electronics Engineers (IEEE) 802.11 series of standards ensured the availability of appropriate networking standards for wireless LANs.

In 1997, the IEEE 802.11 Working Group adopted the first wireless local area network standard, named IEEE 802.11, with data rates up to 2 Mbps [14]. After that standard, IEEE started several Task Groups (designated by the letters a, b, c, etc.) to improve this initial standard. Task Groups a, b and g completed their work by providing relevant extensions that are also known as Wi-Fi (*wireless fidelity*). The 802.11b Task Group produced a standard for WLAN operations in the 2.4 GHz band, with data rates up to 11 Mbps [15]. This standard, published in 1999, has been very successful. In the same period, the 802.11a task group specified a standard for WLAN operations in the 5 GHz band, with data rates up to 54 Mbps [16]. Finally, Task

Group 802.11g has just completed a specification for a higher speed extension of 802.11b [17]. 802.11, 802.11b and 802.11g are all operating in the 2.4 GHz band, and this enables some forms of backward compatibility. On the other hand, 802.11a operates in the 5 GHz band and it cannot interoperate with the other 802.11 standards. In the following, we will provide an overview of the state of the art of the IEEE 802.11 standard family by focusing on the original 802.11 standard and its enhancements (a, b and g). Special attention will be paid to IEEE 802.11b, that is currently the reference technology for wireless networking. In addition, we will also present the activities of Task Group 802.11e that attempts to enhance the MAC with QoS features to support voice and video over 802.11 networks.

14.4.2 Network architectures

IEEE 802.11 WLANs can be implemented via the infrastructure-based, or the ad hoc paradigm. An infrastructure-based architecture relies on the existence of a centralized controller over each cell, often referred to as an *access point (AP) or base station*. The AP controls all the communications within its transmission range, i.e. its *service area*. All mobile devices inside the service area can directly communicate with the AP, while communications among mobile devices must go through the AP. The drawbacks of an infrastructure-based WLAN are the costs and time associated with purchasing and installing the infrastructure. These costs may not be acceptable for dynamic environments, where people and/or vehicles need to be temporarily interconnected in areas without a pre-existing communication infrastructure (e.g., intervehicular and disaster networks), or where the infrastructure cost is not justified (e.g., in-building networks, specific residential community networks, etc.). In these cases, the infrastructure-less or ad hoc networks represent a more efficient solution. An ad hoc network is a peer-to-peer network formed by a set of stations within the range of each other that dynamically configure themselves to set up a temporary network. No fixed controller (e.g. access point) is required, but users' mobile devices are the network, and they must cooperatively provide the functionality that is usually provided by the network infrastructure, e.g. routers, switches, servers and admission controllers [18]. In the IEEE 802.11 standards family, infrastructure-based and ad hoc networks are named the *Basic Service Set (BSS) and Independent Basic Service Set* (IBSS), respectively [1].

The natural choice for building WLAN systems that should efficiently manage broadband and converged services for nomadic users is the infrastructure-based layout. In particular, the BSS can be the basic building block to construct extended WLANs where multiple access points are organized to provide an enlarged area of wireless coverage, as shown in Figure 14.1. The access point behaves as the portal to the core telecommunication networks for the users served by the different BSSs. When the connection to the Internet network is provided by a BWA system, the resulting network infrastructure forms an all-wireless solution to distribute the broadband services to the end users, although they may be spread over a wide geographic area.

14.4.3 High-speed WLANs: 802.11a and 802.11g

There is a continuous demand for increased bandwidth in WLANs. To tackle this issue, IEEE 802 Working Groups defined two standards: 802.11a and 802.11g, both providing a maximum data rate of 54 Mbps.

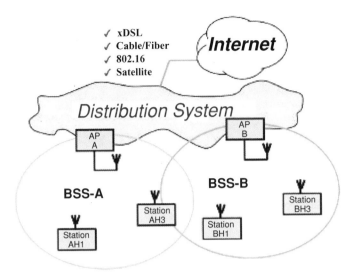

Figure 14.1   An extended basic service set

802.11a provides data rates up to 54 Mbps by operating in the 5 GHz band, which is called the universal networking infrastructure (UNI) band. The UNI band is further divided into three subbands. Devices operating in each subband are subjected to different transmission constraints. For implementing WLANs, only the UNI-1 (5.15–5.25 GHz) and UNI-2 (5.25–5.35 GHz) can be used.[2] Specifically, UNI-1 is reserved for indoor use, while UNI-2 is both indoor and outdoor. Compared to the 802.11b technology, the 802.11a technology guarantees a more efficient frequency reuse, since four nonoverlapping channels exist within each subband, while only (up to) three nonoverlapping channels are available in 802.11b. Therefore, 802.11a-based WLANs can use eight different channels. This larger number of available channels greatly encourages WLANs' deployment. The increased spectral efficiency of the 802.11a systems is due to the use of the orthogonal frequency division multiplexing (OFDM) modulation technique [19]. Specifically, in traditional frequency division multiplexed (FDM) systems, nonoverlapping channels are obtained if the spacing between channels is greater than the symbol rate. Instead, OFDM systems adopt a fixed spacing among the channels because the orthogonality among carriers is achieved by distributing the data over multiple carriers spaced at planned frequencies. Hence, guard bands are eliminated so as to increase the spectral efficiency of the overall system.

An 802.11a-compliant device can transmit using one of the available transmission rates. In particular, a multirate station can operate at the following throughput steps: 6, 9, 12, 18, 24, 36, 48 and 54, where 9 and 18 are optional. However, the throughput increase is obtained at the cost of using modulation techniques that are less robust to interference, thus reducing the allowed transmission ranges. Furthermore, by using the 5 GHz spectrum, the reduction in the transmission ranges is more marked than the corresponding reductions in the 2.4 GHz spectrum [19]. To overcome these limitations, the 802.11g standard has been proposed. Specifically, the

---

[2] UNI-3 is reserved for outdoor bridging.

Table 14.1   The relationship between 802.11b, 802.11g and 802.11a

	Single/Multicarrier	802.11 b	802.11 g	802.11 a
1	Single carrier	Barker	Barker	
2	Single carrier	Barker	Barker	
5.5	Single carrier	CCK	CCK	
6	Multicarrier		OFDM	OFDM
9	Multicarrier		OFDM	OFDM
11	Single carrier	CCK	CCK	
12	Multicarrier		OFDM	OFDM
18	Multicarrier		OFDM	OFDM
24	Multicarrier		OFDM	OFDM
36	Multicarrier		OFDM	OFDM
48	Multicarrier		OFDM	OFDM
54	Multicarrier		OFDM	OFDM

802.11g is interoperable with the 802.11b systems because it operates in the 2.4 GHz band, while offering the higher data rates of the 802.11a standard. In detail, the 802.11g system uses the same modulation technique as the a and b systems, providing a maximum transmission rate of 54 Mbps for WLANs that operate at 2.4 GHz. The 802.11g standard uses OFDM modulation but, for backward compatibility with 802.11b, it also supports complementary code keying (CCK) modulation (see Table 14.1 for a synoptic view of these standards).

14.4.4 Quality of Service in 4G WLANs: the 802.11e standard

The basic 802.11 MAC protocol is the *distributed coordination function* (DCF) that works as a listen-before-talk scheme, based on *carrier sense multiple access* (CSMA). Furthermore, the 802.11 defines a *collision avoidance* (CA) mechanism to reduce the probability of collisions due to two or more stations transmitting concurrently. Since the CSMA/CA protocol is a random access MAC protocol, it does not provide support either to differentiate flows or to guarantee QoS requirements such as delay, jitter, etc. To support time-bounded services, the IEEE 802.11 standard defines the *point coordination function* (PCF) to let stations have priority access to the wireless medium, coordinated by a station called the *point coordinator* (PC). The PCF operates similarly to a polling system: the PC provides (through a polling mechanism) the transmission rights at a single station at a time. There are problems with the PCF that can induce a poor QoS performance, these are leading to the design of enhancements to the MAC protocol discussed in this section. Among many others, it is worth pointing out the unpredictable delays before the beginning of the contention free periods,[3] and the unknown transmission durations of the polled stations.

To support applications with QoS requirements, the IEEE 802.11e WG is developing a supplement to the MAC layer to provide QoS support for LAN applications. It will apply to 802.11 physical standards a, b and g and will add to them the ability to support time-sensitive

---

[3] Up to 4.9 ms in IEEE 802.11a.

multimedia data. Although all the details have not yet been finalized, nevertheless the basic elements have been selected. Specifically, the e standard defines two main access modes that are backward compatible with DCF and PCF:

1. The *enhanced DCF* (EDCF) that extends the DCF to provide service differentiation.
2. The *hybrid coordination function* (HCF) that modifies the PCF to have more efficient polling.[4] This method is under the control of the hybrid coordinator (HC) that is typically located in the access point.

Similarly to 802.11, the channel access is subdivided into contention periods (CPs) and contention free periods (CFPs). EDCF is used during CPs only, while the HCF can operate in both periods.

14.4.4.1 Enhanced DCF

EDCF enhances the DCF access scheme to support service differentiation by introducing, in each station, four access categories (ACs) that are used to support user traffic with eight different priorities. Each AC operates as a *virtual station* following the DCF-like rules with its own parameters for interframe spacing and contention parameters. The values of these parameters are selected to give the highest access probability to the traffic with the highest requirements. As each access category corresponds to a separate virtual station, a station with multiple ACs will have an internal contention among its ACs (besides the access contention with the other stations). Contentions among ACs within the same station are locally managed: the highest AC performs the transmission attempt (that may collide with the other stations), while the other ACs behave as when a collision occurs (virtual collision). As in DCF, a real collision occurs if more stations start transmitting at the same time.

14.4.4.2 Hybrid coordination function

Some types of interactive and synchronous traffic may require QoS guarantees (e.g. bounded delays) more stringent than those provided by EDCF service differentiation. To manage these classes of traffic, the HCF access mode is included in the 802.11e standard. The HCF access mode uses a coordinator (*hybrid coordinator*) generally located in the access point that is in charge of starting, when necessary, controlled access periods in the channel access. The HC performs bandwidth management by allocating the transmission opportunities (TXOPs) to 802.11e stations. A TXOP defines the time interval in which a station can transmit, and it is characterized by a starting time and a maximum duration. The HCF access method operates during both contention periods (CPs) and contention free periods (CFPs) by initiating and managing time intervals during which only the hybrid coordinator and the polled stations can access the channel. During CPs, the hybrid coordinator can start controlled access periods in which bursts of frames are transmitted using the polling-based channel access mechanism. This also provides guaranteed services under heavy load conditions. A signaling protocol is used by HC to learn which stations must be polled. Specifically, the stations signal to the hybrid

---

[4] While PCF is optional, HCF is an 802.11e mandatory mechanism.

coordinator the requirements of their traffic streams (e.g. required bandwidth, maximum delay, etc.) during controlled contention periods that are started on the channel by the HC sending a special control frame. By using this information, the coordinator determines the time instants at which stations must be polled.

14.4.5 Future directions

Several Task Groups are currently active inside the 802.11 Working Group to fix technical and regulatory issues, and to promote the worldwide use of this technology [20]. For example, Task Group 802.11h is working to comply with European regulations for 5 GHz WLANs, while 802.11d is developing an autoconfiguration protocol through which an 802.11 device receives the regulatory information related to the country in which it is operating.

14.4.5.1 Security issues

The activities currently ongoing in Task Group 802.11i deserve special attention. This Task Group is working to improve security in 802.11 networks. Wireless communications obviously generate potential security problems, as an intruder does not need physical access to the traditional wired network to gain access to data communications. The IEEE 802.11 Working Group addressed security issues by defining an optional extension to 802.11: wired equivalent privacy (WEP) [13]. WEP is a form of encryption that should provide privacy comparable to that of a traditional wired network. If the wireless network has information that should be secure, then WEP should be used. WEP supports both data encryption and integrity. The security is based on a 40-bit secret key. The secret key can either be a default key shared by all the devices of a WLAN, or a pair-wise secret key shared only by two communicating devices. Since WEP does not provide any support for the exchange of secret keys, the secret key must be manually installed on each device. Cryptographers have identified many flaws and weaknesses in WEP, pointing out that WEP is only useful for preventing casual traffic capture.

Challenges for wireless security

To address these security issues, the 802.11i Task Group has been set up. It will apply to 802.11 physical standards a, b and g by providing new encryption methods and authentication procedures. IEEE 802.1x forms a key part of 802.11i. 802.1x is based on the IETF Extensible Authentication Protocol (EAP) [21]. The computing power of today's handheld devices can be compared to that of their desktop counterparts of one generation earlier. This phenomenon, while driving more and more functionality into handheld wireless Internet-enabled services, is also driving the security risks that we have today in desktop computing into wireless devices [13]. Consequently, basic security requirements, such as network services availability, entity authentication, data confidentiality, data integrity and nonrepudiation, which have been advocated for Internet-based communication and computing, are directly advocated for wireless e-applications as well. In fact, in addition to presenting the usual Internet security threats in online applications, wireless devices introduce new challenging issues specific to their mobility – the shared communication medium and the lack of interoperability while roaming between different security domains. Additionally, in ad hoc networks, we need to consider also the lack of assistance from a fixed infrastructure and from a central authority [22]. The following are examples of specific, and especially sensitive, requirements for the emerging wireless e-applications: most importantly, guaranteeing confidentiality and integrity of credit card numbers and the secure storage of signatures used for purchases; also, guaranteeing privacy and anonymity of customers who would like to privately retrieve information before making decisions about their purchases; finally, guaranteeing the privacy of bids in auctions.

14.4.5.2 Spectrally efficient WLANs based on picocells

A key problem of existing WLANs is that their spectral efficiency is too low in order to effectively support multimedia traffic. In the recently standardized 2nd generation WLANs, high spectral efficiencies can be obtained by using higher order constellations. Today, the standards for the second generation WLANs already provide a 54 Mbps mode, based on the spectrally efficient 64 QAM modulation scheme. However, shifting towards more efficient modulation techniques reduces the coverage. Hence, for a fixed total coverage area, more access points would need to be installed to cover small-size (e.g., pico) cells [23]. While picocellularization seems to be a natural approach to increasing the spectral efficiency of WLANs, it exhibits some drawbacks affecting the overall system cost. The main relevant drawbacks of picocellularization are:

- *Cell size vs. network reinstallation cost.* Obviously, with a decreasing cell size, an increasing number of APs need to be installed and interconnected in order to maintain a fixed total coverage area. The increasing interconnection effort is especially annoying.
- *Cell size vs. total system capacity.* From a statistical point of view, the total capacity required within a cell tends to increase (in a stochastic sense) with the number of users. In other words, the capacity of a cell with a large number of users can be set more or less according to the average required data rate. However, for a picocell with only a small number of users, the maximal instantaneous data rate is the most relevant design parameter. Hence, for given traffic patterns and assigned total coverage area, the overall capacity of a picocellular network architecture needs to be larger than that of a network with larger cells.
- *Cell size vs. handover and routing.* It is anticipated that future multimedia wireless LANs will consist of fixed and mobile user terminals. Examples of the latter are portable computers, personal digital assistants, cordless phones, etc. Therefore, a mechanism will be required that continuously tracks whether a user terminal crosses a cell boundary, and another that maintains the session continuity when such crossing occurs. In practical systems these mechanisms are implemented via handover and routing protocols, respectively. Obviously, in a picocellularization approach, handover frequently occurs.

14.4.5.3 Spectrally efficient WLAN based on space division multiple access (SDMA)

To avoid the drawbacks of picocellularization, 4G WLAN architectures are emerging where the bandwidth is reused within each cell. This result may be achieved by using advanced APs equipped with antenna arrays able to separate, through signal processing, multiple subscribers transmitting over the same physical channel (e.g. using the same frequency band and time slot). Such multiple access schemes, referred to as SDMA, are effective as long as the subscribers

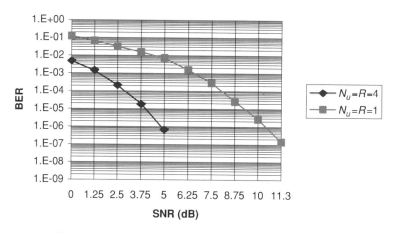

Figure 14.2   Performance of SDMA with MLSE at the AP

are not located in the same angular region with respect to the AP. Specifically, let us consider N_u single-antenna subscribers, operating in the same frequency band, that simultaneously transmit their data sequences to an AP equipped with $R \geq 1$ receive antennas. Thus, the composite signal received at the output on the receive antennas may be modeled as described in [24] and the task of the maximum likelihood sequence estimator (MLSE) present at the AP is to detect all subscribers' messages by elaborating the R received sequences. Under the assumption that the channel gains are reliably estimated at the AP, the MLSE reduces to a minimum-distance sequence estimator that may be recursively implemented via the Viterbi Algorithm (VA) [25]. Figure 14.2 reports the simulated performance of the above detector for the cases $R = N_u = 4$ and $R = N_u = 1$ when the modulation format employed by subscribers is BPSK and the channel gains of the subscribers are Rayleigh-faded. These cases may be representative of a WLAN system with spot radius of 50 m working in the 5 GHz band and supporting a throughput-per-user of about 25 Mbps. Two main remarks may be drawn from Figure 14.2. First, both curves of Figure 14.2 exhibit the same slope, thus confirming that the AP with $R = 4$ receive antennas is able to separate up to $N_u = 4$ subscribers without diversity loss with respect to the benchmark case of $R = N_u = 1$. Secondly, the system with $R = N_u = 4$ exhibits an SNR gain of about 6 dB over the benchmark system with $R = N_u = 1$, that in practice may be translated to an increase of the hotspot coverage radius of about 1.5 times.

The main conclusion is that, in principle, an AP with R receive antennas is able to guarantee frequency reuse factors up to R when suitable signal processing is performed at the AP. Similar results for frequency reuse may be achieved for the downlink (e.g. the broadcast channel from the AP to the subscribers) via optimized beamforming of the signals radiated by the multiple antennas present at the AP [26].

14.5 Wireless backbone and wireless local loop: the IEEE 802.16 standards

Local multipoint distribution systems (LMDS) [27] can be considered the precursors of the current fixed BWA networks for MAN environments. Research and experimental implementations of LMDS-specific architectures [28], [29] support the conclusion that this technology is a viable alternative to cabled access networks such as fiber optic links, coaxial systems using cable modems and xDSL connections over copper wires. However, whether or not the BWA technology is proved to be affordable and feasible, broadband wireless access to television, multimedia, Internet and data services has not yet become widespread. This has been partially due to technical and economical circumstances, partially to a lack of standardization. Standardization is one of the key factors in making a technology appealing, both to industry and to the users' marketplace. Although LMDS systems have been in use in recent years, the development of a standard denotes that the BWA industry is sufficiently mature to enter the mass market. In fact, a standard protects users from being locked into a particular vendor. It also lowers equipment costs and allows interoperability between competitors, fostering a widespread deployment and a successful marketplace penetration.

The IEEE has undertaken standardization of BWA, beginning in July 1999, with the purpose of supporting the development and the deployment of fixed wireless access systems. Specifically, the 802.16 Working Group has been established to address the 'first mile/last mile' connection in wireless metropolitan area networks, working towards LMDS-type architectures for broadband wireless access. The Working Group has completed, and is currently enhancing, two IEEE standards:

- The IEEE 802.16 WirelessMAN™ standard ('Air Interface for Fixed Broadband Wireless Access Systems') [30], released in April 2002. This focuses on the efficient use of bandwidth between 10 and 66 GHz and defines a medium access control (MAC) layer that supports multiple physical layer specifications customized for the frequency band of use.
- The IEEE standard 802.16.2 [31], published in September 2001. This provides a recommended practice on 'Coexistence of Fixed Broadband Wireless Access Systems' covering 10–66 GHz.

The need for reliable non line-of-sight (NLOS) operations, together with the opportunity to expand the system scope to license-exempt bands, has led to the development of an amendment to the 802.16 standard. In January 2003, Task Group a released the Amendment 802.16a ('Medium Access Control Modifications and Additional Physical Layer Specifications for 2–11GHz') that enhances the 802.16 standard to support broadband wireless access at frequencies from 2 to 11 GHz, both in licensed and license-exempt bands (also known as wireless high-speed unlicensed metropolitan area network, or wireless HUMAN). The IEEE 802.16a standard also specifies the MAC enhancements required to support interoperable point-to-multipoint and (optional) mesh topology BWA systems.

14.5.1 Service scenarios and network architectures

The growing demand for high-speed Internet access to small office/home office (SOHO), to small to medium enterprise (SME) premises and to residential customers is increasing the need for network infrastructures able to provide last mile broadband access. Traditionally, broadband and converged services are accessed through large-capacity, high-speed fiber optic networks. Today, small businesses and residential customers typically use wired networks such as cable modem networks and DSL. Currently, wireless short-range high-speed technologies, such as the IEEE 802.11b (Wi-Fi) technology, provide cost-effective solutions to offer customers high-speed remote access to the Internet in public areas, known as hotspots.

Moreover, technological advances promise big increases in access speeds with the 802.11a and 802.11g standards, enabling wireless public networks to play a major role in delivering broadband services and applications to users. However, the mass deployment of broadband services is still far from being a reality, due a number of significant economical and business issues associated with the different techniques outlined above. Specifically, fiber networks are scarcely deployed worldwide, and extending these networks with cable is costly and time consuming. While DSL is already being deployed on a large scale, the copper-based links typically offer 128 kbps to 1.5 Mbps data services, and suffer from critical distance limitations that cause the service not to be available to remote subscribers [12]. Finally, although WLAN-based systems can provide high-speed wireless connections in isolated densely populated 'islands' with the benefit of a rapid and low-cost network deployment, it is still necessary to have a wired connection between the hotspot access point and the Internet backbone to establish a true 'wireless' Internet service. Generally speaking, the overwhelming cost of providing wireless hotspots is due to the fixed and recurring costs of the wired infrastructure.

Considering the deployment trade-offs associated with traditional broadband cabled networks, wireless MANs are attractive and cost-effective alternatives to wired access networks for connecting homes and businesses to core Internet and broadband data services worldwide. Specifically, an 802.16-based infrastructure has the capability to provide high-speed communications services in broad geographic areas without the unaffordable infrastructure costs needed for the deployment of wires to individual sites and locations. The BWA system is also a natural choice for underserved rural and outlying areas with low population densities. Moreover, in these areas, the natural environment adds additional constraints and difficulties to the problem of cable placement, maintenance and upgrade. A two-way satellite transmission system could be a viable BWA choice for rural areas. However, satellite systems tend to have low upstream bandwidth and have limited application in more populous areas due to limited spectrum availability and high latency. On the other hand, IEEE standard 802.16 BWA systems offer true differentiated broadband services at minimal cost. They let thousands of users share capacity for data, voice and video. They are also scalable: carriers can expand them as subscriber demand for bandwidth grows, by adding channels or cells.

In order to discuss the different deployment trade-offs and benefits associated with each of the possible system layouts, in the last part of this section we briefly outline the network architectures defined for 802.16-based BWA systems.

The IEEE 802.16 architecture consists of three components: the base station (BS), the repeater and the subscriber station (SS). The network components are fixed (e.g. not movable). As indicated in the standard specification, the BS provides the connectivity, management and control of the SSs. Moreover, the BS is also connected to external backhaul networks such that it provides connections to public core networks and other stations simultaneously [32]. As shown in Figure 14.3(a), the wireless MAN network employs a point-to-multipoint (PMP) architecture where each BS serves a number of SSs in a particular area. A PMP system is a star-shaped network where each subscriber connects to the same central hub. The BS transmits on a broadcast channel to all the SSs, while the SSs have point-to-point links with the BS. Since only the BSs are connected to the core wired network, the wireless backbone will have a limited number of wired ingress/egress points toward the Internet, thus lessening the cost of the access infrastructure. The idea behind this concept is that the BWA system delivers (in a cost-effective manner) the broadband Internet services to a building, while users inside

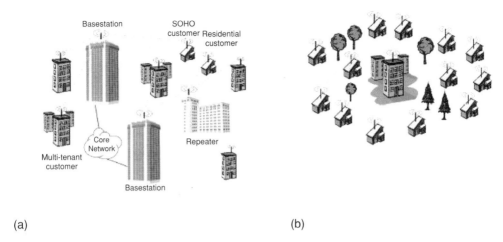

(a)   (b)

Figure 14.3   802.16 wireless MAN architectures. (a) Mandatory PMP topology; (b) optional mesh topology

the building will access the network through traditional local wired networks such as Gigabit Ethernet, or emerging local wireless networks such as Wi-Fi. It is worth pointing out that the design choices for the 802.16 MAC protocol will allow, in principle, delivery of broadband services directly to individual users. This usage scenario will probably require the definition of new physical layers, but the design of the wireless MAN MAC protocol can fully manage the QoS requirements of such a connection.

At the high frequencies (>10 GHz) used in 802.16 systems, line-of-sight (LOS) communications are needed because the system can tolerate a limited amount of multipath interference. However, the LOS condition introduces important issues on the spectrum efficiency and network scalability. Specifically, in typical urban scenarios it is quite difficult to have LOS between SS and BS. The solution is to provide coverage from multiple locations by guaranteeing multiple overlapping BS coverage, but in doing so, the network provider increases the infrastructure cost and reduces the spectrum efficiency. In fact, each SS in an area covered by a BS shares the same channel, hence, the more BSs, the higher the interference between cells and the more channels required to cover an area. To solve these problems, the 802.16 Working Group designed a BWA system (in the IEEE 802.16a standard), which operates at lower frequencies (2–11 GHz), where reliable NLOS operations are feasible. The NLOS communications enable the optional mesh topology, depicted in Figure 14.3(b). While in PMP systems the traffic is sent only between BS and associated SSs, in the mesh mode, the traffic can be routed through other SSs or directly between SSs.

The mesh mode introduces new challenges and complexities in the MAC protocol because nodes have to synchronize with their neighbors (that form the node's neighborhood) and the neighbors of their neighborhood (the two-hop neighbors that form the extended neighborhood). Moreover, the channel access coordination should reach the extended neighborhood, augmenting the MAC complexities.

However, the mesh topology also introduces fundamental benefits over classical PMP systems. Specifically, the mesh solution allows the establishment of connections with visible SSs,

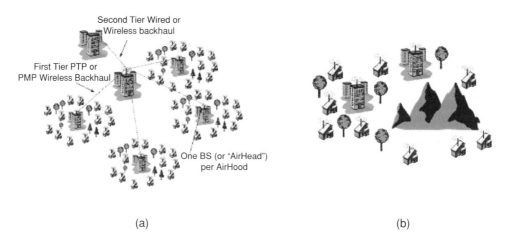

First Tier PTP or PMP Wireless Backhaul

Second Tier Wired or Wireless backhaul

One BS (or "AirHead") per AirHood

(a) (b)

Figure 14.4 Mesh mode applications. (a) Multitier networks; (b) scaling coverage

thus increasing the chance that a new site is served by at least one existing system node. More-over, the mesh concept favors frequency reuse, because the connections are point-to-point links and a single channel can be reused many times in the system. Finally, the use of generally short-length paths and lower power, reducing the path loss and adjacent node interference, allow the adoption of more efficient modulation schemes, thus increasing the link throughputs.

Figure 14.4 exemplifies two applications of mesh-based broadband wireless access systems. In particular, Figure 14.4(a) shows a complex network architecture where the mesh mode is used in overall small areas (referred to as 'airhoods') served by a single BS, while the BSs form a network tier that operates as the distribution system of the backhaul services. We can note that the BS distribution system might itself be a PMP wireless backbone. Figure 14(b) shows a typical situation in rural areas, and points out how the mesh mode could be advantageous for the scaling properties of the network. In fact, the nodes can select the best link among the available ones, further gaining advantage from the reduced interference. A final remark is needed to point out that a two-hop path can achieve a higher throughput than a direct link, because shorter links allow the use of more efficient modulation schemes, thus increasing the system capacity.

14.5.2 Physical layer for wireless backbone

The PHY layer for the 10–66 GHz bands is designated as 'Wireless MAN-SC' in the IEEE 802.16 standard, since it employs a single-carrier modulation to establish LOS communica-tions. The channels used in this PHY environment are typically large, ranging from 20 to 25 MHz (typical US allocation) or 28 MHz (typical European allocation), with raw bit rates up to 134 Mbps. Data bits coming from the MAC layer are randomized, FEC encoded and mapped to a QPSK, 16 QAM (optional for the uplink), or 64 QAM (optional) signal constellation.

The PHY layers for the 2–11 GHz bands are addressed in the IEEE 802.16a standard. The fundamental design choices are driven by the consideration that the receiver has to deal with significant multipath interference due to the NLOS communications [33]. Three PHY layers have been specified in the standard documents:

- Wireless MAC-SCa. This PHY layer is based on single carrier technology.
- Wireless MAN-OFDM. This PHY layer uses orthogonal frequency division multiplexing with 256 subcarriers.
- Wireless MAN-OFDMA. This PHY layer uses orthogonal frequency division multiple access with 2048 subcarriers. In the OFDMA mode, the active carriers are divided into a subset of carriers, termed subchannels, which may be addressed to different (groups of) individual receivers. This solution can offer advanced multiplexing in tiered MANs and support selective multicast applications.

The parameters of the downlink and uplink transmissions, such as modulation type, forward error correction code, guard times, etc., can be adaptively selected by SSs, effectively balancing different data rates and link quality. The modulation method may be adjusted almost instantaneously, i.e. frame-by-frame, for the optimum data transfer.

14.5.3 Future directions

The 802.16 standards have specified four different PHY layers, each of which must operate in a wide range of different frequency bands. Moreover, a large number of system options and features are available. The Working Group will continue to expand the standard to cover additional markets, by releasing new amendments to the standards. However, this plethora of options hinders the manufacturers in implementing fully compliant products. To address this problem, amendment c to the IEEE 802.16 standard [34] was published at the end of 2002. This document aimed to detail the system profiles, and to list sets of features and functions to be used in typical implementation cases. However, the scope of this amendment is limited to 10–66 GHz systems. Moreover, the 802.16 Working Group does not address the creation of test specification, because the standards primarily concentrate on system requirements. Therefore, the industry has set up the WiMAX Forum [35], a testing and branding organization modeled after the Wi-Fi Alliance. The WiMAX intends to foster a more competitive BWA marketplace by specifying minimum air interface performance between various vendors' products and certifying products that meet those performance benchmarks.

As indicated previously, the 802.16 Working Group will continue to investigate improvements to the standard, with the purpose of fulfilling the needs of new market sectors and increasing the market penetration of broadband wireless access solutions [36]. To this end, Task Group e was formed in December 2002, aimed at taking advantage of the inherent mobility of wireless media. Specifically, the scope of this project is the definition of enhancements to the IEEE 802.16/802.16a standards to support subscriber stations moving at vehicular speeds, specifying a system for combined fixed and mobile broadband wireless access. This amendment should include functions to support higher layer handoff between base stations or sectors, and extended OFDMA techniques robust to the fading due to mobility. However, the 802.16e Working Group is not the only IEEE Working Group turning its attention to mobile broadband. In fact, other projects are also being formulated, and the Mobile Broadband Wireless Access

(MBWA) Study Group was formed in March 2002. Despite the fact that 802.16e and 802.20 standards will both specify new mobile air interfaces for wireless broadband, there are some important differences between them. For one, 802.16e will add mobility in the 2 GHz to 6 GHz licensed bands, while 802.20 aims for operation in licensed bands below 3.5 GHz. Moreover, 802.16e is looking at the mobile user walking around with a PDA or laptop, while 802.20 will address high-speed mobility issues (speeds up to 250 kilometers per hour). More importantly, the 802.16e specification will be based on an existing standard (802.16a), while 802.20 is starting from scratch. Both the Working Groups are still in a preliminary stage and no public specifications have been released yet.

14.5.3.1 Multi-antenna architectures for wireless backbones

The use of multiple antennas at the terminals of BWA networks, in combination with signal processing and coding, is a promising means to provide high-data-rate and high-quality wireless access for wireless backbones at almost wireline quality [24]. Note that in fixed BWA, as opposed to mobile cellular communications, the use of multiple antennas at the subscriber site is also possible.

The leverages provided by the use of multiple antennas at both the base station and subscribers may be summarized as follows [24, 36]:

- *Array gain.* Multiple antennas can coherently combine signals to increase the signal-to-noise ratio and hence increase the system coverage. Coherent combining can be employed at both the transmitter and receiver and it requires channel knowledge. Since channel knowledge is hard to obtain at the transmitter, array gain is more likely to be available at the receiver [37].
- *Diversity gain.* Spatial diversity through multiple antennas can be used to combat fading and significantly improve the link reliability, thus somewhat relaxing the above-mentioned line-of-sight requirement. Diversity gain can be obtained at both the transmitter and receiver. Some recently developed space–time codes [10, 37] realize transmit diversity gain without knowing the channel in the transmitter.
- *Interference suppression.* In the uplink, multiple antennas can be used to suppress co-channel interference and hence increase the cellular capacity.
- *Multiplexing gain.* The use of multiple antennas at both the transmitter and receiver allows the opening up of parallel spatial data pipes within the same bandwidth, which leads to a linear (in the number of antennas) increase in data rate per user [5, 26, 38, 39].

Summarizing, the use of multiple antennas can improve the capacity and reliability of wireless backbones. With regard to reliability, two of the major impairments of systems are fading and interference from other users. Diversity provides the receiver with several (ideally independent) replicas of the transmitted signal and is therefore a powerful means to combat fading and interference. In recent years, use of spatial (or antenna) diversity has become increasingly popular. Space–time coding (transmit diversity) is a method that yields diversity gain without channel knowledge at the transmitter by coding across both antennas (space) and time [24]. To demonstrate the impact of diversity gain on the performance of a BWA system, Figure 14.5 shows the signal level at the receiver, with and without antenna diversity. We can clearly see that the deep fades in the single-antenna case vanish in the multiple-antenna

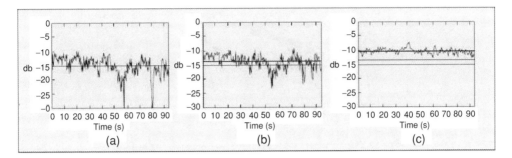

Figure 14.5   (a) Single-antenna case; (b) interference rejection due to multiple antennas (case 1); (c) case 2

one. Hence, diversity makes the channel less fading, which is of fundamental importance for wireless backbones, where deep fades can occur and the channel is changing slowly, causing the fades to persist over a long period of time. Figure 14.5(b) and Figure 14.5(c) show the signal and interference levels for a single-antenna system and a multi-antenna system, respectively. We can see that in the multi-antenna case, the signal level is higher and fluctuates much less. Hence, more aggressive frequency reuse is possible in the multi-antenna case.

14.6 Satellite access and services

Local access via satellite provides another possible alternative for BWA that may be appealing for providing Internet access to sparsely populated areas [40].

14.6.1 Geosynchronous Earth orbit (GEO) satellites and video broadcasting

Satellite services have been available for many years, based on GEO satellites. These last have been mainly used for telephone communications, (analog) television distribution and various military applications. Satellite access clearly has significant advantages in terms of rapid deployment and national coverage, but has high cost and system capacity limitations, especially for the uplink traffic. In fact, although a GEO-based bidirectional Internet service referred to as Direct-PC is currently ongoing in the USA, the capacity of the uplink channel is limited to 12 Mbps [41]. Therefore, up to now, direct video broadcast satellite (DVBS) is the main application of GEO-based systems. The DVBS system utilizes GEO satellites working in the Ku band (e.g., 12.2/12.7 GHz in downlink/uplink, respectively). The architecture of the DVBS access system is sketched in Figure 14.6.

The uplink stream is typically composed of 500 MPEG-coded multiplexed video channels, and the power radiated in the downlink by the GEO transponder is of the order of 120 Watts, to allow the utilization of a small-size antenna at the customer premises. However, while their coverage is very broad, GEO satellite systems present several significant limitations. Power constraints and antenna size (limited to up to 2 ft diameter for residential applications) constrain the aggregated downlink throughput from a GEO satellite to about 100–200 Mbps.

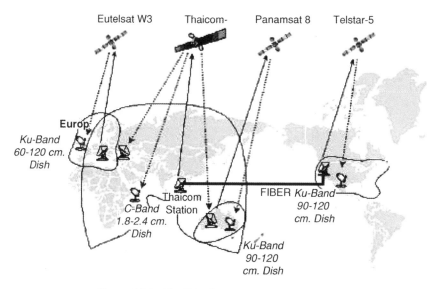

Figure 14.6 The DVBS access system architecture

Although statistical multiplexing effects allow this capacity to be shared over more users than is suggested by simply dividing the aggregated throughput by the peak load per user, nevertheless, the total number of residents that can be served per satellite is limited. New frequency bands are being planned to increase the overall system capacity. In fact, spot beams and on-board switching would provide roughly a factor of 10 improvement in the capacity, at the expense of both reduced geographic coverage and heavier and costlier payload. However, from a market perspective, even though GEO satellites may have a limited total capacity for broadband on a national scale, they may occupy a long-term market niche if the user demand is still limited to a small set of (mostly rural) subscribers. Table 14.2 summarizes the main features of existing GEO-based systems and services.

14.6.2 Towards broadband low Earth orbit (LEO) satellites

Since several GEO systems are already on, the key question is: will GEO systems support BWA? Unfortunately, several technological features limit GEO system's capability to provide broadband access [41]. First, GEO satellite systems suffer from a high round-trip delay of the

Table 14.2 Some current GEO systems and services

Type of system	Frequency bands	Applications	Type and size of terminal	Examples
Mobile GEO satellite	L and S	Voice and low-speed data to mobile UMTS terminals	Laptop/mobile antennas	ImmarSAT, AMSE
Broadband GEO	Ka and Ku	Internet access, voice, video, data	Small-size fixed antennas	Spaceway, Cyberstar, Astrolink

Table 14.3 Implemented LEO systems and related technical features

	Globalstar	Iridium	Skybridge	Teledesic
Services	Telephony	Telephony	High-speed data	High-speed data
Downlink bandwidth	2.483–2.500 GHz	1.616–1.625 GHz	12 GHz	19.3–19.6 GHz
Uplink bandwidth	1.610–1.625 GHz	1.616–1.625 GHz	14 GHz	29.1–29.4 GHz
Modulation scheme	QPSK	QPSK/CDMA	N/D	N/D
Switching	Bent-pipe	On-satellite	Bent-pipe	On-satellite
Number of satellites	56	72	80	288
Maximum throughput per session	9600 b/s	4800 b/s	TBD	16 kb/s (voice) 2 Mb/s (data)

order of 500 ms. This limitation caused some to conclude that GEO satellites are fully useless for data purposes. Although for TCP/IP-based Internet access this delay may be noticeable, it does not seriously degrade the experienced QoS, as long as the end node terminals are properly set up. However, for other applications, this satellite delay is a more serious issue. For example, for Internet telephony, the long delays cause a real degradation of the offered service, since the maximum allowed round-trip delay in a conversation is limited to 200 ms [40].

An alternative that has received a great deal of attention over the past several years is to use LEO satellites sited at an altitude varying from 780 km (small LEO satellites) to 14 000 km (large LEO satellite). A LEO satellite offers a footprint diameter ranging from 4000 km to 6000 km with round-trip delays limited to 12 ms. Furthermore, the utilization of such low orbits enables direct connections between LEO satellites and terrestrial mobile terminals. However, LEO satellites do not occupy constant positions in assigned orbital slots, and we need to rely on multiple satellites to provide adequate coverage. Since power limitations of LEO satellites are less serious than those with GEO satellites, LEO satellite technology, while challenging, can be fielded, as the pioneering Iridium and Globalstar deployments have proved (see Table 14.3). With regard to switching, LEO systems adopt two approaches [40]. In the 'bent-pipe' approach (implemented by Globalstar and Skybridge) the switching operations are carried out on the terrestrial unit, while in the Iridium and Teledesic architectures, the switching operations are carried out onboard the satellite. However, it is a common opinion that within the next few years, the feasibility of LEO satellites for mass-market BWA will be constrained more by economic considerations than technological challenges. In fact, a LEO satellite system requires the launching of many satellites, because in their low Earth orbits, the satellites are in rapid motion overhead, and there must be enough of them to ensure that one is always in range. This means that the system has a very high initial cost, which, in turn, implies that there must be a significant user pool to justify the investment. On the other hand, BWA and LEO satellites may exhibit an appealing performance vs. price trade-off for subscribers sited in sparsely populated (e.g. rural, mountain) areas, that may justify the requested initial investment.

14.7 Summary: future BWA roadmap and diverse technology landscape

Although BWA systems come in multiple flavors, from the outset, some general considerations about possible future developments may be drawn. First, a variety of technology options are possible, depending on the planned customer density. Systems equipped with very few cabled access points have higher per-subscriber costs. However, installing multiple antennas on existing APs allows the leveraging of past investment in access points and related site costs. Furthermore, since little or no cabled infrastructure needs to be installed, deployment may proceed quite fast, but the capacity for adding new customers is limited.

In any case, current BWA system capacity is limited by the amount of radio spectrum available for broadband services (e.g. the fraction of the spectrum allotted via government spectrum licensing polices to broadband services) and the extent to which clever system design can increase the performance achievable from a given amount of bandwidth. As already pointed out, a possible option for increasing performance is to resort to picocellular clusters, so as to increase the frequency reuse. The need to deploy more transceivers (and install wireline to connect them to the provider's network) makes the improvement costly, so that picocellular systems have costs dominated by labor costs approaching those of cabled systems. As already pointed out, an emerging option is to exploit space–time processing techniques combined with multi-antenna transceiver architectures, which hold the promise of roughly a factor of 10 improvement in the user throughput and/or frequency reuse. These emerging techniques rely heavily on improvements in the processing power of application-specific integrated circuits (ASICs) or digital signal processors (DSPs), enabling the real-time signal processing to be performed in affordable hardware.

This contribution supports the conclusion that even though today's 4G technology may be still immature, it is relatively safe to predict that there will be a growing demand for BWA to portable and mobile devices. This is because of fundamental market trends toward smaller personal computing and communication devices (e.g palmtops, PDAs) that, in the long term, are likely to account for the majority of end user devices, in contrast to the fast-growing minority that they represent today. The authors' opinion is that, once untethered computing devices become ubiquitous, today's PC-centric access network (HFC, DSL, and so on) will have to evolve toward hybrid wired and wireless networks, while the last mile, last 100 m, or even last 10 m will be mostly wireless. The cost and performance of broadband wireless networks will thus be crucial to the user's overall experience.

In this context, it is worth noting that there are still significant challenges associated with delivering true broadband services, as defined in this chapter, to mobile devices. Research and development (R&D) challenges faced by developers of '4G' wireless standards include higher aggregate rates (up to 100 Mbps), maintenance of service quality under mobile fading conditions, integration of mobile and fixed network architectures and greater spectral efficiency, capacity and scalability.

The examined different wireless broadband technology options are different in detail. Furthermore, 802.16-based systems may be appealing for supporting market entry by providers that have no access to existing wireline assets. On the contrary, in the most remote areas, a small fraction of users may best be served by LEO satellites, where the very high fixed cost of launching satellites is offset by the very low per-passing costs. Finally, 4G WLANs represent an important example of how a solution to a specific problem – the wireless extension of Ethernet – could become the basis to provide ubiquitous IP-compliant broadband wireless access to mobile users. However, four main technical challenges, namely: ease of use, security, mobility and network management, are still to be overcome before accomplishing this task. Furthermore, the economic viability of the public hotspots market is still unclear.

The final conclusion is that significant research investment will be needed to reach the scalability, cost and performance levels suitable for ubiquitous mobile/portable BWA deployment. Supportive spectrum regulation policies encouraging efficient spectrum usage and easier access to new frequency bands will also be needed to drive this emerging scenario forward.

References

1. Conti, M. (2004), Wireless Communications and Pervasive Technologies, in D.Cook and S.K.Das (eds), *Environments: Technologies, Protocols and Applications,* John Wiley & Sons.
2. IEEE (2002), *Part 15.1: Wireless Medium Access Control (MAC) and Physical Layer (PHY) Specifications for Wireless Personal Area Networks (WPANs),* Standard 802.15.1-2000.
3. IEEE (2003), *Part 15.3: Wireless Medium Access Control (MAC) and Physical Layer (PHY) Specifications for High Rate Wireless Personal Area Networks (WPANs),* Standard 802.15.3-2003.
4. Callaway, E., Gorday, P., Hester, L., Gutierrez, J. A., Naeve, M., Heile, B. and Bahal, V. (2002), Home networking with IEEE 802.15.4: a developing standard for low-rate wireless personal area networks, IEEE Communications Magazine, **August**, 70–77.
5. Bisdikian, C. (2001), An Overview of the Bluetooth Wireless Technology, *IEEE Communications Magazine*, **39**, (12), 86–94.
6. Web site of the Bluetooth Special Interest Group: http://www.bluetooth.com/.
7. European funded UCAN (Ultra wideband Concepts for Ad hoc Networks) project, IST program (IST-2001-32710), Deliverable number: D42-1.
8. Di Benedetto, M.G. and Giancola, G. (2004), *Understanding Ultra Wide Band Radio Fundamentals,* Prentice Hall (in press).
9. Baccarelli, E., Biagi, M., Pelizzoni, C. and Bellotti, P. (2004), A Novel Multi-antenna Impulse Radio UWB Transceiver for Broadband High-Throughput 4G WLANs, to appear in *IEEE Communications Letters*; also available at http://infocom.uniroma1.it/ enzobac/mimouwb1.pdf.
10. Tarokh, V., Seshadri, N. and Calderbank, A.R. (1998), Space–Time Codes for High Data Rate Wireless Communication: Performance Criterion and Code Construction, *IEEE Transactions on Information Theory*, **44**(3), 744–765.
11. Polydoros, A., Rautio, J., Razzano, G., Bogucka, H., Ragazzi, D., Dallas, P. I., Mammela, A., Benedix, M., Lobeira M., and Agarossi, L. (2003), Wind-flex: developing a novel test-bed for exploring flexible radio concepts in indoor environments, *IEEE Communications Magazine*, **41** (7), 116–122.
12. Baccarelli, E., Fasano, A., and Biagi, M. (2002), Novel Efficient Bit-Loading Algorithms for Peak-Energy Limited ADSL-type Multi-Carrier Systems, *IEEE Transactions On Signal Processing*, **50** (5), 1237–1247.
13. Karygiannis, T. and Owens, L. (2002), *Wireless Network Security: 802.11, Bluetooth and Handheld Devices,* National Institute of Standards and Technology (NIST), USA.
14. ANSI/IEEE (1999), *Wireless LAN Medium Access Control (MAC) and Physical Layer (PHY) Specifications,* Standard 802.11-1999.
15. IEEE (2001), *Wireless LAN Medium Access Control (MAC) and Physical Layer (PHY) specifications: Higher-Speed Physical Layer Extension in the 2.4 GHz Band,* Standard 802.11b-2001, Supplement to Standard 802.11-1999.
16. IEEE (1999), *Wireless LAN Medium Access Control (MAC) and Physical Layer (PHY) specifications: High-speed Physical Layer in the 5 GHz Band,* Standard 802.11a-1999, Supplement to Standard 802.11-1999.
17. IEEE (2003), *Wireless LAN Medium Access Control (MAC) and Physical Layer (PHY) specifications Amendment 4: Further Higher Data Rate Extension in the 2.4 GHz Band,* Standard 802.11g-2003, Amendment to Standard 802.11-1999.
18. Razzano, G. and Curcio, A. (2003), Performance comparison of three call admission algorithms in a wireless ad hoc network, in *Proceedings of International Conference On Communications*, ICCT2003, **2**, 1332–1336.
19. Kapp, S. (2002), 802.11a More Bandwidth without the Wires, *IEEE Internet Computing, July–August* 75–79.
20. Varshney, U. (2003), The Status and Future of 802.11-Based WLANs, *IEEE Computer*, **June**, 102–105.
21. PPP Extensible Authentication Protocol (EAP), RFC 2284 (1998) ftp://ftp.rfc-editor.org/in-notes/rfc2284.txt.
22. Zhou, L. and Haas, Z.J. (1999), Securing Ad Hoc Networks, *IEEE Networks*, **13** (6), 24–30.

23. Chiani, M., Dardari, D., Zanella, A. and Andrisano, O. (1998), Service availability of broadband wireless networks for indoor multimedia at millimeter waves, *URSI International Symposium*, Pisa, Italy, pp. 29–33.

24. Paulraj, A., Nabor, R., and Gore, D. (2003), *Introduction to Space–Time Wireless Communications*, Cambridge University Press.

25. Verdu, S. (1998), *Multi–user Detection*, Cambridge University Press, New York.

26. Baccarelli, E. and Biagi, M. (2004), Power-allocation Policy and Optimized Design of Multiple-Antenna Systems with Imperfect Channel Estimation, *IEEE Transactions on Vehicular Technology*, **53** (1), 136–145.

27. Nordbotten, A. (2000), LMDS Systems and their Application, *IEEE Communications Magazine*, **38** (6), 150–154.

28. Mähönen, P., Saarinen, T., Shelby, Z. and Munöz, L. (2001), Wireless Internet over LMDS: Architecture and Experimental Implementation, *IEEE Communications Magazine*, **39** (5), 126–132.

29. Izadpanah, H. (2001), A Millimeter-Wave Broadband Wireless Access Technology Demonstrator for the Next-Generation Internet Network Reach Extension, *IEEE Communications Magazine*, **39** (9), 140–145.

30. IEEE (2002), *IEEE Standard for Local and Metropolitan Area Networks – Part 16: Air Interface for Fixed Broadband Wireless Access Systems*, Standard 802. (9) 16-2001.

31. IEEE (2001), *IEEE Recommended Practice for Local and Metropolitan Area Networks Coexistence of Fixed Broadband Wireless Access Systems*, Standard 802. 16.2-2001.

32. Eklund, C., Marks, R.B., Stanwood K.L., and Wang, S. (2002), IEEE Standard 802.16: A Technical Overview of the WirelessMAN^{TM} Air Interface for Broadband Wireless Access, *IEEE Communications Magazine*, **40** (6), 98–107.

33. Koffman, I. and Roman, V. (2002), Broadband and Wireless Access Solutions Based on OFDM Access in IEEE 802.16, *IEEE Communications Magazine*, **40** (4), 96–103.

34. IEEE (2002), *IEEE Standard for Local and Metropolitan Area Networks – Part 16: Air Interface for Fixed Broadband Wireless Access Systems – Amendment 1: Detailed System Profiles for 10–66 GHz*, Standard 802.16c-2002.

35. Web site of the WiMAX Forum, http://www.wimaxforum.org/home/.

36. Bölcskei, H., Paulraj, A.J., Hari, K.V.S., Nabar, R.U. and Lu, W.W. (2001), Fixed Broadband Wireless Access: State of the Art, Challenges, and Future Directions, *IEEE Communications Magazine*, **39** (1), 100–108.

37. Baccarelli, E. and Biagi, M. (2004), Performance and Optimized Design of Space–Time Codes for MIMO Wireless Systems with Imperfect Channel Estimates, *IEEE Transactions On Signal Processing*, to appear October 2004; also available at http://infocom.uniroma1.it/~enzobac/mimost.pdf.

38. Paulraj, A.J. and Kailath, T. (1994), *Increasing Capacity in Wireless Broadcast Systems Using Distributed Transmission/Directional Reception*, US Patent, no. 5.345.599.

39. Foschini, G.J. (1996), Layered Space–time Architecture for Wireless Communication in a Fading Environment When Using Multi-Element Antennas, *Bell Labs. Technical Journal*, Autumn 41–59.

40. Ibnkahla, M., Rahman, Q.M., Sulyman, A.I., Al-Asady, H.A., Yuan, J. and Safwat, A. (2004), High-Speed Satellite Mobile Communications: Technologies and Challenges, in *Proceedings of the IEEE*, February, 312–339.

41. Hu, Y. and Li, V.O.K. (2004), Satellite-Based Internet: A Tutorial, *IEEE Communications Magazine*, March, 154–162.

Part Four

Case Studies

15

Community Case Studies in North America

Priya Shetty[1] and Ashwin Gumaste[2]
[1]*The University of Texas, Dallas, USA*
[2]*Fujitsu Laboratories, Dallas, USA*

15.1 Introduction

In the chapters so far we have seen multiple technology, business and social aspects of community networks. This chapter focuses on implementation of these concepts in North America. Community networking in North America is a fast growing phenomenon that has attracted a lot of attention (from consumers), as well as controversy from the incumbent carriers. Community networks for many are seen as a larger than life phenomenon, realizing the very basic tenets of absolute liberty such as e-governance and e-democracy. From a business perspective, community networking has the added responsibility of coming forth as a business model alternative to the traditional service providers. Moreover, the community network model provides a stage for demonstration of new technologies and concepts that otherwise would take several years for incumbents to even consider.

Apart from the technological advantages seen in the community network concept, another clear advantage is in terms of the business model. The business model that outlines the community network case has the potential to create an entirely new and more effective business case, as compared to conventional telco networks.

The years after the bursting of the telecommunication bubble witnessed a surge of activity and deployment in the community network space. However, each deployment in the community space can be seen to be unique, and depends on user needs, geographic profile of the community, demographic, etc. Apart from these quantifiable factors, there are also a host of other issues that affect the design and deployment of community networks, such as social hindrances, legal issues and policies of the land that determine the network deployment.

Broadband Services: Business Models and Technologies for Community Networks. Edited by I. Chlamtac, A. Gumaste and C. Szabó © 2005 John Wiley & Sons, Ltd. ISBN 0-470-02248-5.

In this chapter we shall study some deployment case studies in North America, primarily those that are unique and have somewhat different solutions. It should still be noted that each community network design and deployment is unique, tailored to meet its own specific requirement and, hence, worth exploring. At the time of writing, around 145 community networks were planned across the United States alone, and no two community networks had much in common.

This chapter first focuses on some design issues for community networks and then studies case studies discretely.

15.2 Issues affecting community network design

In this section, we shall study some of the issues that affect the design of community networks. Broadly speaking, there are three major areas that require attention: technology considerations, business considerations and societal considerations. The effect that a community network yields on mankind is the complete summation of all the three issues – technological, business and societal impact. It is the last issue that is unique and of paramount importance in driving the concept of community networks as a premier solution for broadband communication.

15.2.1 Technology considerations

In this section we shall discuss some of the technology considerations that affect the community network. Since individual technologies are dealt with in earlier chapters, we will consider technologies from a very broad perspective. There are two broad technology solutions that bifurcate the community network technology space: the LAN technologies and legacy technologies [1].

LAN technologies evolved more recently and were proposed as next generation data-centric solutions. The primary motivation of LAN technologies was the emergence of data communication and IP traffic as the dominant communication methodology, thanks to the World Wide Web. IP traffic and data communication has a unique distribution type – being highly bursty in nature. The burstiness signifies that predicting the amount of data that will arrive is somewhat difficult. As compared to voice communication, which is modeled as a Poisson distribution, data communication is inherently modeled by a more prolific distribution called the Pareto distribution. The distribution results in what is known as long-range variance of the data that flows through, meaning that the arrival of packets is related over long intervals of time. The burstiness is then explained by a phenomenon called heavy tailing of the flow. LAN technologies are primarily Ethernet dominated, as Ethernet is a method to statistically multiplex and frame any data flow whose distribution may be uncertain. Ethernet traffic was first shown in the late 1970s and today it is estimated that 90 % of the world's LANs run native Ethernet technology. The migration of Ethernet from the LAN to the core of networks, typically the metro area where the community space operates, was natural and only a question of time. The emergence of standards that facilitated high-speed Ethernet technologies, such as Gigabit Ethernet and eventually 10 Gigabit Ethernet, led to this migratory path of enhancing Ethernet communication from the LAN area to the metro. In addition to the standardization process, there was also a push from component vendors to create a portfolio of technologies that facilitated the movement of LAN technologies into the metro, and hence community, area. This push included innovations such as Gigabit interface converters, or GBICs. GBICs allow conversion of Gigabit Ethernet frames from electrical to optical media, and hence enhance the reach of Ethernet technology from LAN to MAN.

A second technology solution is the TDM approach, comprising legacy technologies like SONET/SDH and ATM services. TDM basically means time division multiplexing of signals in order to achieve efficient bandwidth utilization. In TDM, multiple smaller speed signals are combined to create a high-speed signal. The combination is done by time division multiplexing of the individual signals such that each signal is periodically sampled and combined into a higher speed signal. TDM is particularly efficient for transport of signals that are quality sensitive, such as latency requirement signals like voice communication. In addition, TDM provides excellent fairness to all its users by evenly distributing the channel bandwidth. TDM technologies include, but are not limited to, SONET/SDH, ATM, frame relay and all the associated slower speed tributaries.

15.2.2 Business considerations

To achieve a successful community business model, business considerations are of primary importance. This means that the community model has to take into account considerations such as the products and services it will offer to its inhabitants, in addition to the return on investment (ROI) plan that it will have to showcase to its investors.

15.2.2.1 Product and service portfolio

The product and service portfolio determines which products and services the community can offer to its inhabitants. Generically, community management will do a detailed study that demographically identifies the key services that are essential to the inhabitants. It will also identify growth points and services that are expected to grow. This forecasting has to be accurate, else it can lead to catastrophic financial problems. Key services such as voice and data, while producing high volumes of traffic, provide little revenue, and hence should be observed somewhat conservatively. Services such as video programming, Internet gaming and multimedia are emerging products that need more attention in the years to come. Emerging communities look at these services for enhancing the ROI model, as well as improving the lifestyle of the populace.

15.2.2.2 ROI considerations

Community design must consider ROI models. Service-provider-like models fail when it comes to design and deployment of community networks. The service provider model is a profit-centric one, while the CN model is a more social model. The aim is not pure profit but to provide services and to provide a mechanism to bridge the digital divide. That is, to enable inhabitants of a community to be able to access information that they otherwise might not be able to access due to their financial position.

15.2.3 Social issues

A community network, particularly in North America, is prone to certain social issues and we list these briefly. Legal frameworks determine the rights of a community to build and operate a network. Many times, an incumbent has paid high volumes of money to get a license to sell regulated services. Legally, it then becomes difficult for the community to compete. Fair agreements need to be developed for relations between the community and the incumbent. Right of way (ROW) issues are also of paramount importance, and most community network deployments face competition from the incumbent service providers.

15.2.3.1 Incumbent dealings

To the incumbent provider a CN is competition. Not just competition but a competitor who prices his products far lower than the incumbent. The incumbent does have a tendency to dampen a community effort. Such cases have been recorded in Chapter 4.

15.2.3.2 BOT schemes

An impressive scheme by which the community can create a network is the BOT scheme. The BOT scheme stands for build, operate and transfer. The community management invites tenders from external companies to build the network. The external company then can operate the network for a fixed amount of time, thus recovering the building costs. Eventually, the external company transfers the network back to the community. This kind of scheme involves almost zero investment from the community and enhances the lifestyles of the inhabitants by providing them with a state-of-the-art network with extremely useful services at a price that is both affordable to the inhabitants and regulated by the community management.

In the next section, we will review some case studies using the concepts mentioned above. These case studies are adapted from the webblog: http://communityfiber.blogspot.com/ [2].

15.3 Case studies

15.3.1 Case study 1: Douglas County School System, Georgia

Our first case study takes a look at the high-capacity fiber network and the wireless wide area broadband network implemented by the Douglas County School System in Georgia, in conjunction with NetPlanner Systems Inc. of Norcross, Georgia and Navini Networks of Richardson, Texas. NetPlanner Systems Inc. assisted with the Siemon-certified cabling infrastructure installation for the fiber network, while the wide area wireless network was overlaid on the wired network using non line-of-sight technology from the Ripwave range of products from Navini Networks. The network is illustrated in Figure 15.1.

Douglas County reflects the growing breed of municipalities that would like to control their own networks and be able to offer services to their government agencies and constituencies that the local incumbent provider is not offering, such as fiber-to-the-home (FTTH). The Douglas School System faced upgrade costs to the tune of millions of dollars when they recently required an upgrade to their old communications network, comprising a 1.5 Mbps leased system on the standard T1 frame relay backbone connection, in addition to recurring charges for services related to the network. The administrators of the school system teamed up with NetPlanner Systems Inc. and Navini Networks to allow the deployment of a self-owned

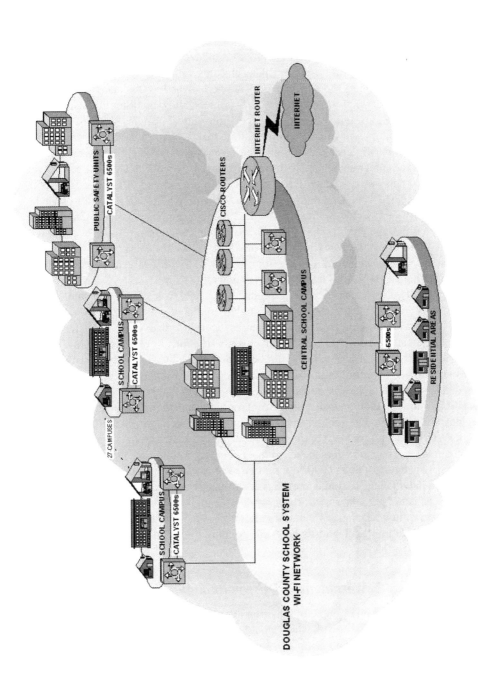

Figure 15.1 Douglas County School System Wi- Fi network

fiber backbone and a wide area wireless network, which enabled a cost reduction of more than $ 500 000 a year, with an exceptional improvement in the network performance.

The new 10 Gbps network was implemented over 18 months for a total cost of $2.2 million (US), eliminating $320 000 in recurring data communication charges incurred by its earlier 1.5 Mbps leased system from their local BellSouth provider. Installations by NetPlanner deployed Siemon Category 6 solutions, fiber RIC enclosures with three to ten managed Siemon RS2 racks per school. Countywide, this network supports over 15 000 data ports which provide connectivity to over 7000 computers and 1500 printers, allowing faster file sharing, faster Internet connections and resulting in immense economic and educational benefits.

Greystone Power partnered with the school to allow space on their power poles for the fiber. The Douglas County network has been designed based on Cisco AVVID solutions (architecture for voice, video and integrated data), with the core driven by a Cisco GSR (Gigabit switch router), a 12 000 carrier class route, the distribution layer made up of Cisco 6513s, 6509s and 3550 Gigabit catalyst switches and access layers driven by Cisco 2950 catalyst switches. This fiber optic network would eventually link up 28 school campuses, county administrative and judicial facilities and public safety departments within the county. It has also allowed the replacement of older technologies with newer ones, for instance the conversion of the Douglas County School System telephony system from the PBX/ Centrex system to Voice over IP (VoIP), facilitating a complete IP telephony environment.

The Navini Networks Ripwave range of products was chosen to deploy the wireless wide area network to create a more conducive learning environment which would extend beyond the school walls, without being hindered by expenses incurred with expensive installations and unscalable broadband networks. The Navini Networks solution allows for a non line-of-sight deployment, which is ideal for the highly timbered area of Douglas County, based on multicarrier synchronous beamforming (MCSB) technology. The Ripwave solution comprises a fully redundant base station, an element management system and modems. The base station can easily fit into a rack or enclosure and uses patented adaptive phased array beamforming technology with a small and light antenna. The Ripwave EMS is an IP-based element management system, which provides comprehensive element, subscriber, bandwidth and service management, in addition to fault management, performance monitoring and security and configuration management for up to 1000 base stations. The low-cost product does not require any hardware installation at the customer premises and offers a range of up to eight miles from the base station and high-speed data rates on a par with cable modems and DSL. This Navini-based wireless network allows Douglas County families access to several facilities, individualized instructional programming, diagnostic and skill building tools and a school-filtered Internet portal, to name but a few.

15.3.2 Case study 2: Washington's DC-Net through MCI

Our second case study is an example of a metropolitan area network, DC-Net in Washington DC, a high-speed fiber optic network which would connect more than 350 buildings, including structures housing government offices, public safety personnel, general city offices and schools. This citywide network with a backbone of 2.5 Gbps, would be supported by two network operation centers run by systems integrator Science Applications International; MCI would be the service provider and the network hardware would be provided by Avaya and Cisco. The network is illustrated in Figure 15.2.

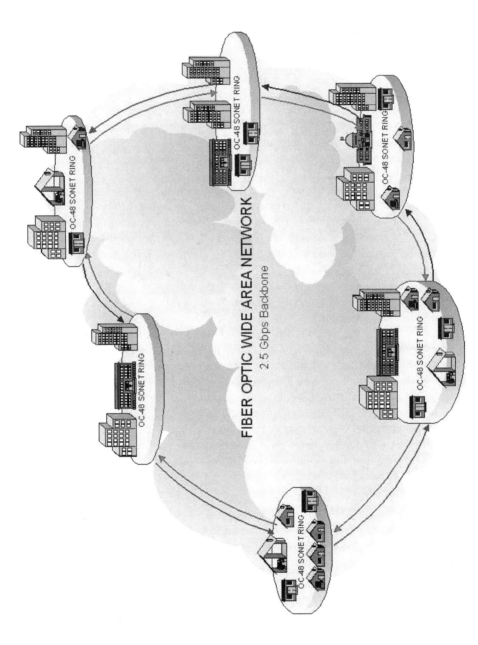

Figure 15.2 DC-Net: metropolitan area network in Washington DC

Prior to building DC-Net, the city was networked through a maze of T1 and T3 lines, leased by various departments from Verizon at a cost of around $30 million a year. Each of the city departments managed its own network. The main purpose of city officials building an integrated citywide network which would consolidate its voice and data traffic, was to have a reliable infrastructure for public safety communication and various other forms of data and voice communication. The $93 million dollar implementation is expected to yield significant savings of up to $10 million dollars a year.

DC-Net comprises several OC-48 SONET rings; SONET was chosen primarily due to its excellent protection mechanism, owing to its self-healing property of connection restoration within 50 ms in the event of a fiber cut. The network uses Avaya's PBX system to support around 50 000 ISDN handsets and Cisco's optical provisioning platforms, routers and switches. Part of the fiber used for building the $93 million network was purchased from Comcast, Level 3 Communications and Starpower, while city officials deployed the remaining sections with self-owned fiber optic lines.

DC-Net is an excellent example showcasing how a community network provides a common platform to coalesce several different types of management platforms, protocols and technologies, thereby increasing the different types of service which could be offered by the network. DC-Net includes network management platforms designed by multiple vendors, including HP OpenView, and software from Hummingbird, InfoVista, Micromuse and Remedy.

The network is capable of supporting multiple network protocols, including Ethernet, frame relay, ISDN, multiprotocol label switching, switched multi Megabit data service, TCP/IP and TDM. The network has allowed the introduction of new innovative applications, such as real-time interactive video for the public school system, routing 911 emergency calls to improve emergency response times and transferring data from industrial automation systems for applications such as traffic light monitoring. Among several other objectives, city officials envision offering a website where citizens would be provided with updates regarding weather conditions in real time, by real-time analysis of the data gathered through DC-Net. DC-Net would eventually also support an independent private high-speed wireless data network for public safety communications, and another network for a unified communications center for handling emergency calls. Plans to develop and support such innovative services on DC-Net will make the government-owned community network one of the most advanced private metropolitan area networks of its kind.

15.3.3 Case study 3: broadband power line system in Cape Girardeau, Missouri

This case study is an example of a community network providing high-speed broadband access based on power line communications (PLC) technology. Regional CLEC (competitive local exchange carrier) Big River Telephone Company has plans to deploy a full scale BPL (broadband over power line) service in Cape Girardeau, Missouri by partnering with the St. Louis based Ameren Energy Communications and Main.net Communications. The electric utility Ameren would provide the distribution framework, by allowing Internet data transmission through its power lines, thus providing the much needed alternative to the expensive DSL and cable net connections, and the upgrade from the lower dial-up speeds.

Big River provides customer support, technical assistance and a link from the Internet backbone to Ameren's power lines and servers for email access to the customers. Compared to wireless and fiber optic broadband solutions, PLC technology is a relatively lesser known alternative for providing high-speed Internet access. Ameren has conducted rigorous tests with its BPL service in Cape Girardeau over the past eighteen months, and an average test customer can currently download images, video and data at speeds equivalent to DSL services. Ameren and Big River plan to offer BPL at rates which would be competitive with the existing DSL service offered by SBC Communications and the cable service offered by Charter Communications Inc., at around $30 to $50 a month.

The biggest advantage for the movement to broadband access through power line communication is the omnipresence of power lines. With every home and business in the nation tied to the electric power grid, community networks deployed using PLC will have the advantage of lower installation costs and times, as opposed to fiber optic and wireless solutions. Since electricity is more prevalent in homes than cable or DSL lines, PLC brings out a new possibility for a vast communication infrastructure, bringing Internet access to the most remote or rural areas as well, where broadband access has lagged. The Federal Communications Commission has been a strong proponent in favor of power line access technology to bring high-speed Internet access to as many homes as possible.

15.3.4 Case study 4: wireless broadband community network for City of Oceanside, California

Our fourth case study takes a look at the wireless broadband community network being deployed for the City of Oceanside, California through consultations with Sun Wireless, a local value-added reseller (VAR), and with equipment supplied by Proxim Corporation, a leading player in wireless networking equipment.

The various departments, community agencies and residential areas of The City of Oceanside had been connected through expensive DS3 and T1 leased lines, like many other cities in North America. A considerable growth in the community resulted in a subsequent increase in the demands for bandwidth and speeds from the existing network. The city required a more cost effective network, which would meet the increasing bandwidth requirements and accommodate the growing community, while providing higher levels of performance, quality and reliability, thereby making cost reduction, reliability and scalability the primary criteria for the deployment of the network. The costs incurred in the purchase or lease of fiber and the time required for installing the fiber backbone are much higher than deploying a wireless solution. Proxim's end-to-end wireless solution was selected because of the relative ease of installation of its components, faster speed of installation and lower costs. The capacity of the Proxim wireless network can be easily extended to accommodate future growth, which fulfilled one of the main motivations behind deploying the network.

The Tsunami range of wireless products from Proxim has been used in the deployment of this metro area wireless network, which connects Oceanside's major facilities, public safety departments and utility departments, such as the water and harbor facilities. The initial deployment covered a smaller geographical area, linking six sites with 45 Mbps Tsunami point-to-point radios. Its successful deployment resulted in an extension of the original network to include another 15 sites, with 100 Mbps radios in the backbone and the relocation of the 45 Mbps radios to the newer sites. The network comprises several repeater sites using four feet tall antennas over a 12 mile hop radius. The network has allowed significant cost savings of up to $150 000 a year by eliminating leased line fees and lowering network operating costs.

15.3.5 Case study 5: citywide wireless network for Mount Pleasant, Michigan

Last Mile Technologies based in Mount Pleasant, Michigan teamed up with Proxim Corporation to deploy the first citywide Wi-Fi network for residential users and local business units in Mount Pleasant, Michigan, see Figure 15.3.

Prior to the deployment of the network, students going off the campus of Central Michigan University (CMU) would have to rely on the dial-up speeds of the university modems, which could provide only a basic level of connectivity unsuitable for high-bandwidth applications. There was thus an increasing demand for a cost-effective high-speed Internet access solution which could be extended beyond the university campus. A network offering such ubiquitous access to services such as video-on-demand (VoD), Voice over IP (VoIP), interactive streaming video and high-speed Internet access was required to extend the learning environment of the university. Since the solution was required to provide ubiquitous access at an affordable price, the network was deployed in the apartment complexes surrounding the campus by creating a consolidated infrastructure for the old and new complexes using the existing T1 wired backbone of the old apartments.

Using the fiber backbone of Last Mile Technologies, the citywide wireless network has been deployed through Proxim's wireless Tsunami point-to-point and point-to-multipoint outdoor wireless products. In order to maximize the cost effectiveness of the solution, the high-speed wired backbone of the existing T1 lines has been extended to all the major units within the complexes, and the network has been further extended through Proxim's outdoor wireless products to provide coverage to a larger area of the city. Residential users can connect to the Ethernet network through any of the 802.11b access points of the indoor wireless LAN. The entire wireless network has been deployed at a cost of around $240 000 and is expected to yield around $290 000 in terms of revenue per year. The wireless backhaul is a 45 Mbps point-to-point link, and the network comprises several 20 Mbps and 60 Mbps base station units and 20 Mbps subscriber units located in different sectors within the city.

The network covers a four to six mile radius, along with student housing, to provide high-speed broadband access to around 4500 students of Central Michigan University (CMU) residing in eight apartment complexes around the university. With around 60 percent of the university town housing being rental, there are plans to extend the network capacity to accommodate around 5000 additional students who are expected to reside in additional apartment complexes. High-speed wireless Internet access plans are available at $24.95 a month for speeds of 256 kbps to 512 kbps, and at $49.95 a month for speeds from 256 kbps to 1.5 Mbps, primarily targeted at local businesses. Around twenty local business units comprising manufacturers, architects, gas stations and city offices have already joined the network, in addition to hotels and apartment complexes in other areas of the city. The network provides enterprise-class speeds and capacity to businesses in a way that has not been possible before, and allows for easier expansion to accommodate an increased subscriber base in the event of future growth.

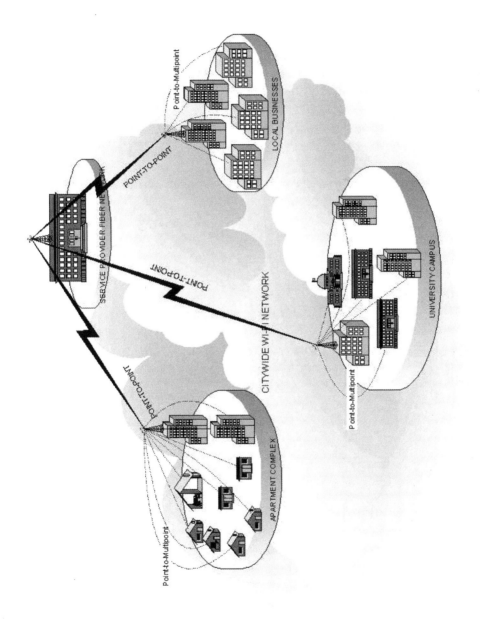

Figure 15.3 Wireless wide area network in Mount Pleasant, Michigan

15.4 Summary

In this chapter, we have showcased case studies for community networks. Community networks in North America are becoming prolific and a true alternative for providing broadband communication to residential and business users. We first discussed some of the insights into community network development and then considered the case studies. Five case studies were discussed.

References

1. Gumaste, A. and Antony, T. (2004), *First Mile Access Networks and Enabling Technologies*, Pearson Education Press.
2. http://communityfiber.blogspot.com/.

16

European Broadband Initiatives with Public Participation

Csaba A. Szabó

Budapest University of Technology and Economics, Budapest, Hungary

16.1 Introduction

This chapter is a selection of characteristic examples of broadband projects created by some form of public participation in different countries in Europe. The case studies differ by the business models, products and services provided, the models for public participation and the underlying technology.

The first example is Stockholm municipality's dark fiber network. Although many metropolitan fiber networks have been built worldwide, it is not typical that the project becomes profitable after five years, as in the Stokab case.

In the next example, FastWeb is an alternative Italian broadband telecommunications company offering Internet, voice, broadcast TV and video-on-demand services over a single connection. In Milan, the city where it started, it utilizes the dark fiber network infrastructure created by Metroweb, a joint venture with the local utility AEM.

The telecom infrastructure deployed by Endesa in some Spanish cities like Zaragoza or Barcelona is based on different technology, namely on power line communications (PLC). Field trials have proven the feasibility of building PLC-based community network infrastructures.

Another technology alternative to fiber is satellite, which has been proven as a viable solution in the South West region of Ireland.

Finally, a special form of public participation is *demand aggregation* which, according to the UK experience outlined in this chapter, seems to be an efficient way to solve the public sector's broadband needs by aggregating the demands and consolidating the physical networks into a single purchase.

Broadband Services: Business Models and Technologies for Community Networks. Edited by I. Chlamtac,
A. Gumaste and C. Szabó © 2005 John Wiley & Sons, Ltd. ISBN 0-470-02248-5.

16.2 Stokab's dark fiber metro net in the Stockholm region – a profitable project in business terms

16.2.1 Overview

A high visibility project in Europe is the Stockholm example. The government of the city created an optical metropolitan infrastructure and is selling the capacities to telecom and other companies. The main product the city is selling through a publicly owned company, *Stokab*, is dark fiber. Stokab, 100 % owned by the city of Stockholm, owns the network and is also the entity contracting customers. In addition, Stokab also operates a transport network for data and telecommunications to the City of Stockholm and its own enterprises and companies. This is the only instance in which Stokab plays the role of an operator, otherwise it is operator and competition-neutral, providing fiber optic connections to all service providers at equal terms.

The Stokab network covers most of central Stockholm, and is expected to cover Stockholm County with an area of about 6500 km². 5600 km of cable was laid down, with over 1 200 000 fiber kilometers [1], see Figure 16.1.

Stokab is now entering a new phase of development, focusing not only on continued extension of the network but also on ensuring optimal use of the infrastructure. This is being carried out in collaboration with various stakeholders, such as property owners, operators and service providers. An example of such cooperation is the municipal housing corporation

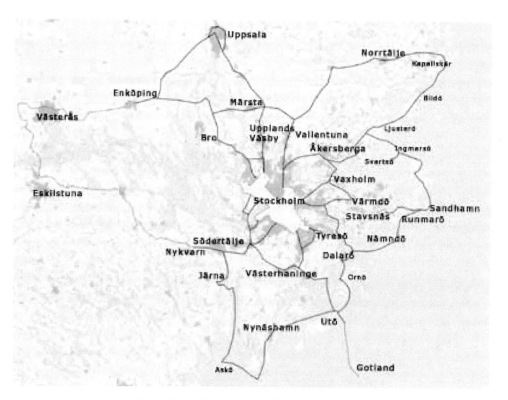

Figure 16.1 The coverage of the Stokab network

Svenska Botstaeder's 'Fiber-to-the-home' project, in which 43 600 residential and some 10000 commercial tenants will be provided with broadband access.

16.2.2 Connection provisioning and pricing

The customer can create its own network in point-to-point, ring or star topology by leasing several connections with the same and/or different delivery points. Fiber optic connections can be provided in quantities from a single pair and upwards, but only in pairs. Redundant connections can be built via different routings. A fully redundant connection provides different routing, even including the access part. The connection is normally provided at a delivery point in the block or property where it is possible for Stokab to establish the network. Long-distance connections may require amplification along the routing and upon request, Stokab can provide space for repeater equipment to do this.

Public prices are available for fiber optic connections up to a connection length of 20 km; for longer connections, special quotes are prepared. The price structure consists of a one-time fee, which depends on the location and leasing fee (to be paid quarterly), depending on the distance. For example, quarterly leasing fees for 1 km, 5 km and 10 km fiber pairs are SEK 7 928, 19 133 and 32 142, respectively [2]. (The SEK to EUR ratio was 9:1 in June 2004.)

16.2.3 Collocation and location services

Stokab's unique collocation service is called Meeting Point, which is a new type of marketplace where Stokab's customers enjoy a unique opportunity to meet at one location [2]. Stokab has several Meeting Points strategically placed throughout the Stockholm region and on the island of Gotland. Currently, a number of industries are represented at Meeting Points: operators, telehouse companies, banks, IT companies, medical companies, and media companies, as well as municipalities and county councils. A company connected to Meeting Point can quickly and conveniently make new contacts with other, already established companies.

Meeting Point is unique because it is operated by an independent party. The Meeting Point marketplace enjoys a very high level of security, including protection against unauthorized access, alarm monitoring, reserve power, emergency service, and the possibility to obtain 24-hour access.

Stokab also provides sites for radio antennae on water reservoirs and other tall buildings within Stockholm County. These will be linked to Stokab's other fiber optic networks in order to create a sound infrastructure for GSM, UMTS, LMDS, etc.

16.2.4 Customers

16.2.4.1 Operators

Stokab has approximately 60 operators as customers. They can be broken down into the following primary groups: telecom operators, Internet operators, cable TV companies, mobile telephony operators and network capacity operators. As an alternative to constructing their own networks, these operators lease fiber in Stokab's network in the desired topologies. Dark fiber may be regarded as a 'raw material' in the operator's product range and imposes no limits on the services which may be offered.

16.2.4.2 End users

Companies requiring high data transfer capacity and efficient and secure communications can lease 'their own' fiber optic connections from Stokab. Because the end customers are responsible for all communications equipment, they benefit from freedom of choice and complete control. This also means that it is easier to influence and change transfer speed and communications technology, etc. Stokab's end customers include banks, insurance companies, retailers, media companies, university colleges, urban networks, property owners, computer and IT companies and others.

16.2.4.3 Municipalities

With its 'own' urban network, a municipality can control its own IT infrastructure as well as its costs. By linking its operations, such as schools, administrative bodies and municipal companies, it is possible to maintain cost-efficient communications and increase capacity in the network without making new investments in infrastructure. By maintaining a common access point to the network, several different service providers can offer services to municipal operations. This creates healthy competition leading to quality improvements and lower prices. Several of the municipalities in Stockholm County have chosen to maintain their 'own' network within Stokab's network. Consequently, the municipal network can take advantage of connections to other urban networks.

16.3 FastWeb – a new generation of telecommunications networks and services

Today FastWeb – or e.Biscom after the recent merger with its holding company – provides telephony, Internet and television services to over 400 000 homes in 13 Italian cities, including Milan, Rome, Turin, Venice and Naples. In Milan, where it first started, it covers about 25 % of the city's residents.

The FastWeb case is quite unique in different respects. First, it started by creating a metropolitan fiber optic infrastructure based on a public–private partnership, in a similar way to many community networks worldwide. The company Metroweb was a dark fiber only provider, with the participation of AEM, the power utility company of the city of Milan. The business model of separating the infrastructure company from the service provider company still exists, Metroweb now belongs 100 % to AEM, whereas now FastWeb belongs 100 % to e.Biscom.

Secondly, it became a fully fledged voice/video/data service provider in many major Italian cities and represents the biggest competitive telco to the incumbent Telecom Italia. Its shareholder mix has changed, now the (public) utility AEM has a minority shareholding in e.Biscom of about 12 %.

Lastly, it is also an example of transborder penetration, not typical in Europe. e.Biscom bought and helped to build HanseNet, a large community network in the Hamburg area, in partnership with Hamburg's municipal power utility. e-Biscom developed a similar service portfolio to that of FastWeb. Subsequently, e-Biscom sold its shares in HanseNet to Telecom Italia, thus helping the latter, e.Biscom's biggest competitor in Italy, to penetrate into Germany

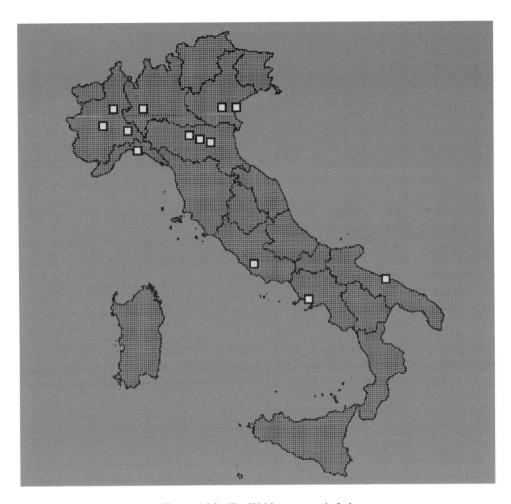

Figure 16.2   FastWeb's presence in Italy

[3] and allowing e.Biscom to concentrate on its core Italian business. Figure 16.2 illustrates the presence of FastWeb in Italy.

e.Biscom started at the right place, in Milan, Italy and the right time, in 1999. Regarding the place, the major Italian cities do not have cable TV operators, so the business objective to build an FTTH network for providing broadband services was viable. The timing was also excellent: in 1999, during the dotcom boom, it was not difficult to obtain financing, thus e.Biscom was able to build and extend its infrastructure at a quick pace. The business case was promising, not only because of the lack of CATV, but also due to the fact that the incumbent, Telecom Italia, was just about to start with ADSL, the only 'broadband' facility to compete with.

As for the technology, FastWeb operates Ethernet networks over fiber, the simplest and least costly networking option in metropolitan areas. Their universal platform for data/voice and video services is IP. Equipment manufacturers include Alcatel, Marconi and Siemens (for DSLAMs) (DSLAMS are for DSL) and Cisco (for Ethernet over fiber).

In addition to FTTH, FastWeb also uses DSL technology in many areas, providing DSL customers with the same services due to the fact that FastWeb's DSL goes up to 6 Mbps. The reason that FastWeb connects more customers with DSL is due to DSL technology advancements (the customer is detached from Telecom at the local switch and connected to FastWeb's network). IT can also be more cost-effective, depending on population and business density. Wherever FastWeb is using ADSL, it leases a small part of the infrastructure from the incumbent (copper wire) only from the local switch, the crucial part of the network is in fiber and belongs to FastWeb. In Italy, the technical conditions in the subscriber loop are particularly suitable for ADSL because the average length of the last mile is short and the copper wire is of good quality.

FastWeb was the first company in the world to offer broadcast TV over DSL. Today video over xDSL is no longer unique in Europe, several incumbent telcos offer, or are planning to offer, 'triple play' (voice/video/data) services over xDSL as this is the most natural way for them to enter into these new services, thus securing new revenue streams. Analysts say that, in the medium term, xDSL will remain the dominant broadband technology, with market share well over 90 %, and even in Italy, the current 15 % share of FTTH will drop to 10 % [4].

Services provided by e.Biscom's FastWeb include both residential and business services:

- telephony (Voice over IP), billed according to several different schemes, including a flat rate with unlimited traffic;
- data: broadband Internet access, always-on, e.g. according to an unlimited traffic scheme;
- video broadcast: 80 to 120 TV channels over FTTH (over Ethernet) and over DSL. A particularly attractive program is the popular Italian League soccer coverage;
- video-on-demand: over 5000 titles;
- virtual video recording;
- videoconferencing for businesses (and videocalling on the TV set at home);
- VPNs (virtual private networks) for businesses.

Prices for residential services are € 85 per month for limitless domestic landline telephony and limitless 10 Mbps Internet, another € 10 for 80 TV channels and video-on-demand, with a choice of over 5000 titles, some of which are free of charge, and movies for € 3 to 6 per movie [5].

As for the business performance, FastWeb became EBITDA positive two years after it started to build the infrastructure. What is also important to note is their impressive ARPU (average revenue per user), which is about € 900 per year, the highest in Europe. The operator claims that it is three times higher than the ARPU of a traditional telecom operator [6]. For key financial figures, see Figure 16.3.

16.4 Community networks based on PLC: the Endesa field trials in Spain

16.4.1 Why PLC?

Power line communications (PLC) is a proven technology, meaning transmission of data using segments of the power distribution networks, mainly the low voltage grid, thus offering an alternative to traditional telecommunication access networks. PLC technology can also be

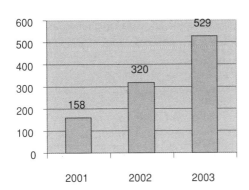

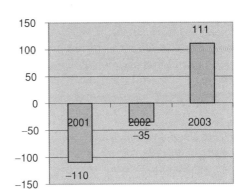

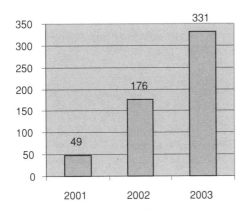

Figure 16.3 FastWeb's key financial figures

successfully used in the medium voltage segment of the electricity network, thus serving as a distribution level of the telecom network.

Electricity utility companies, as well as product manufacturers, are ready for mass deployments and commercial roll outs, after carrying out several successful pilot projects. In this area, Europe has been playing a leading role, but American and Asian manufacturers are quickly catching up.

The key strength of the PLC technology is that it uses existing infrastructure. To put it simply, PLC transforms the conventional electric plug into a connection point for telecom services. Due to this fact, PLC-based networks can be rolled out more rapidly than any competitive technology. New generation PLC equipment provides broadband rates several times better than the current xDSL technologies. Chip designers and manufacturers are working on enhanced versions of their products that will provide data rates up to 200 Mbps and include additional functionalities. The high bandwidth makes it possible to provide not only Internet access, but also telephony (Voice over IP) and audiovisual and multimedia services (i.e. video-on-demand, videoconferencing, etc.). Additional important applications, such as remote control, surveillance or remote meter reading, can be implemented on PLC as well, which means a high-value PLC application for the utility [7].

16.4.2 Network technology using PLC

PLC technology is illustrated in Figure 16.4. Two areas where PLC is used are shown in the figure, the low voltage part, which can serve as the access part of the telecommunication network, and the medium voltage part, which can serve as the distribution network. As for the access part, head ends are installed in power substations, repeaters are installed in the meter room of the building and the intra-building power lines connect terminals, located in customer flats. It is also straightforward to use the in-house electric network for building a home network, thus providing a PLC signal in all rooms of the customer's house. The medium voltage power network can connect different low voltage substations, thus serving as a distribution network.

The complete network architecture, including the service provider's network, is shown in Figure 16.5. In the access part of the network, PLC is exclusively used. In the distribution part, a mixed solution can be used. Medium voltage PLC can be combined with fiber optics and with other technologies, such as xDSL or LMDS.

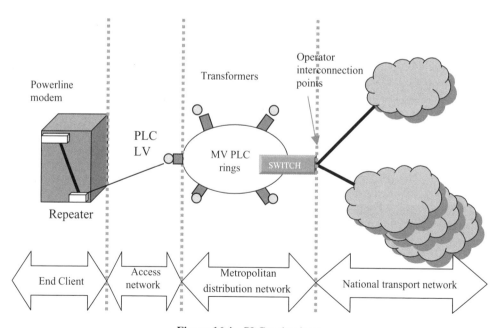

Figure 16.4   PLC technology

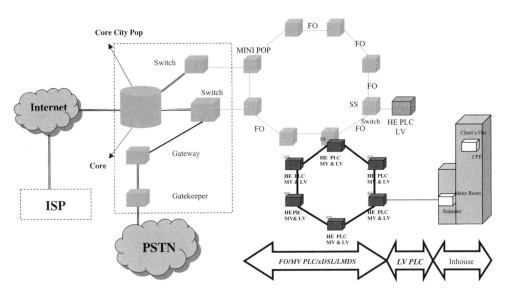

Figure 16.5 PLC network architecture

16.4.3 Business models

Ease and speed of deployment of PLC is one of the reasons why it can be considered one of the most competitive technologies. Also, its economics is favorable, its capex is competitive with other technologies as the wired network is already in place. Experience shows that the capex needed is about the same as for ADSL. In the future, further improvement is expected as the equipment costs will decrease due to mass production.

PLC can be an important way to foster competition in the local loop. When most countries lifted the monopoly of the incumbent operators, local loop unbundling was thought to be a way to introduce competition in the local area. However, it did not go to plan, so far allowing only very limited competition. At the same time, competition in the local area is very important for the penetration of broadband services, and PLC can be viewed as an alternative network infrastructure that introduces facilities-based competition in the local area.

There are two main PLC business models, with some flavors added to them in particular cases: 'the wholesaler model' and 'the retail model' [7].

According to the wholesaler model, the utility plays the role of an access operator whose customers are mainly telcos and ISPs. Normally, it is an open network on an equal basis to all telcos interested in a given area. The utility is in charge of deployment and operation and maintenance of the infrastructure based on PLC. The advantage of this model for the utility is that it can focus on its key strengths and core competence. This model is particularly useful for telecom service providers that do not own network infrastructure in the access parts. The utility can build such networks for the service providers in areas which they want to penetrate based on their market forecasts.

The retail model means that the utility plays the role of the telecom service provider itself. The utility not only owns and operates the network infrastructure, but it also takes care of the necessary sales and marketing activities, as well as the billing for the services provided. It is important to emphasize that in this case, the utility is competing with all telcos that are active in that area.

16.4.4 The mass field trial in Zaragoza

Endesa, one of the largest private electricity companies worldwide, with more than 22 million customers in Europe, North Africa and Latin America, has been active in the development of PLC systems. It started with pilot trials in Barcelona and Seville in 2000 and Santiago (Chile) in 2001, then Endesa carried out mass field trials in Zaragoza during 2001–2003, and in 2004 its efforts were focused on the commercialization of services in Zaragoza and Barcelona.

The PLC network architecture in Zaragoza is similar to that of Figure 16.5. In the backbone part, leased dark fiber is used, while the distribution part was built using over 80 medium voltage PLC links. The 140 low voltage transformers, serving as head ends for the access links, were connected partly with optical cable and partly with medium voltage PLC links. The total number of buildings was 300 and the total number of users (households) connected was 2103 during the five months of the deployment, thus the rate of connecting users was 25 users per week on average [8]. The total coverage of homes was over 20 000.

Services provided via the pilot network were Internet access and telephony. 87 % of users used the Internet daily. Many benefited from the broadband capabilities: 20 % of users communicated more than 100 Mbytes daily (the Spanish average being around 5 Mbytes). Voice traffic showed heavy usage, especially after opening up the network for long-distance calls. User satisfaction surveys showed that most users were very satisfied with the quality of the Internet service; they found it better than the ADSL service (4.32 for PLC vs. 3.65 for ADSL). The general satisfaction with the telephone service was good, an average 3.65 as compared with 3.99 obtained for Telefonica, the incumbent service provider.

The results of the massive field trail can be summarized as follows [8].

- applicability of both low and medium voltage power grids;
- provision of broadband Internet access and VoIP services;
- commercial availability of PLC equipment provided by leading manufacturers;
- viability of the interconnection with the public Internet and telephone networks;
- cost-competitiveness of PLC against ADSL;
- positive experience in installation of PLC networks (no affecting electricity services);
- satisfactory results of electromagnetic interference measurements and no official complaints received.

Endesa is currently commercializing PLC services in Zaragoza and Barcelona. The business model is a wholesale services model. The potential market in Zaragoza is the area covered by the existing deployment of the field trial, while in Barcelona, a new area was selected and the PLC network was deployed.

Services provided are Internet access at data rates 128, 300 and 600 kbps. Telephony services are provided using VoIP technology. Different service packages were configured by the telecom operator in order to be competitive with the other technologies offered in those cities. The market penetration in the commercial area is higher than the average penetration rate for ADSL in Spain. More than half of the customers have contracted VoIP.

16.5 Broadband to rural areas via satellite – the South West Broadband Programme in Ireland [9]

The South West region is one of the eight regions in Ireland, which includes Cork and Kerry counties. Its territory is about 12 000 km², with a population of 580 000, about half of it living in rural areas. The region is a well-developed one, with a high industry presence, this being mostly electronics and chemical. The GDP in the region is 20 % higher than the EU average, and the unemployment rate is circa 4 %. All this means a strong regional economy, but the core of business is located in the greater Crok City area, with a decline of agricultural, fishing and traditional industries. Therefore, there is real need to revitalize the rural economy by creating new growth opportunities and fostering the development of indigenous SMEs, smaller towns and villages throughout the region [10].

The South West Regional Authority (SWRA) the is statutory body responsible for the coordinated delivery of all the public services in the region. Over the last seven years, the authority has been involved in a number of programs using information technology to promote delivery of services and a balanced regional development, and foster a vibrant knowledge-based economy.

One of the main stumbling blocks repeatedly encountered was the lack of broadband availability in more remote towns and rural communities. Due to the low population density and high installation costs, it is often very difficult to convince telecommunications providers to provide adequate broadband capacity, by way of fiber in the ground, in many areas throughout the region. It may not be considered economically viable to do so, as the demand may not match the costs involved.

The Regional Authority concluded, after consultations with a number of organizations in the region who are committed to securing adequate broadband connections throughout Cork and Kerry, that *satellite broadband* could be an answer. The ARTES Program, funded by the European Space Agency (ESA), provided a real opportunity to test the water using a number of different technologies and applications.

ARTES is the ESA/Industry partnership program for telecommunication projects. The objectives are to identify and carry out research, development and demonstration activities in the space segment, Earth segment and communications services areas on the basis of proposals made by industry, in order to improve the match of developed technologies and services with industry's and businesses' own future exploitation activities.

The fundamental objectives of the SWB Programme were:

- to prove that satellite can deliver the required bandwidth to support a range of applications which have, to date, been either completely prevented or seriously constrained due to lack of broadband bandwidth;
- to demonstrate the usability of satellite-based technology and applications for the public sector, business development and rural communities;
- to evaluate the interconnectivity with the IEEE 802.11b/g standard-based wireless LANs as access solutions;
- to have satellite-based broadband formally adopted into regional planning policy;
- to support and sustain innovation in technology and business development;
- to evaluate and disseminate experiences for the benefit of other peripheral regions in Ireland and abroad.

Satellite communication systems can offer a reliable, always-on, scalable, immediate and cost-effective alternative to traditional terrestrial methods of connecting to the Internet. This is especially the case when matched with wireless local area networks (WLANs) or other last mile solutions.

The overall level of service the user can expect is defined by the level of coverage, usually referred to as the 'footprint'. This is especially important in the South and South West of Ireland; if the footprint is weak, the user can expect problems on a regular basis associated with rain, fog, etc. Generally, problems will manifest themselves as upload problems; meaning that there will be a large variance in the ability to transmit data; not generally in receiving data. By selecting satellites offering the best footprint over this part of Europe, the network availability can be better than 99.7 %. However, across the 16 trials undertaken under SWB, weather conditions did not impact on the general performance of the satellite systems.

The price quoted to the customer is based on three factors[1]:

1. *Data throughput* – the amount of data the customer puts through the satellite in a given month. Typically, most operators will quote a flat monthly rate which is unmetered, and the qualification that there is a 'fair usage policy'. This is not unusual, but the customer needs to fully understand what is allowed and what is not, and if necessary, opt for a more expensive usage option.

2. *Maximum burst speed* – the desired access speeds the customer would like to see when connecting to the Internet. It should be remembered that satellite access is on a 'shared' basis, meaning that although the stated maximum burst speed is 256 kbps, the typical access speed could be significantly lower depending on the number of other users sharing a given channel.

3. *Contention ratio* – the most important factor in defining the price the customer pays per month. In order to make satellite access affordable, satellite access is shared amongst a number of users. The level of over-subscription is referred to as the contention ratio, and is effectively a measure of the level of difficulty associated with getting service. Each time the user tries to send a web request via satellite, his/her terminal will first listen to see if it is clear to transmit. If not, the terminal will 'back-off' for a random number of milliseconds and then try again. If there is too much contention on the channel, then clearly, the user's terminal will have to back-off so often that the effective transmission rate will be quite slow. One way that satellite companies can offer very cheap rates is to increase the contention ratio. It can happen with heavily contended systems that some individual users will not be able to gain access for considerable periods. A lot of companies have put a lot of work into finding the optimal contention ratio, the perfect number that will offer low cost to the customer while still offering an excellent degree of service.

The SWB Project worked with a range of statellite industry partners and wireless providers to test a variety of systems and packages in a number of settings – i.e., business, community (WLAN), public sector, education and health sectors.[2]

---

[1] Taken from www.swra.ie/broadband, written by Ildana Teoranta www.ildana.ie [11]

[2] SWB partners – West Cork Adult Learning Centre (West Cork VEC), Cork County Council, Kerry County Council, Udaras na Gaeltachta, ICS Skills, Post Graduate Medical and Dental Board (Cork University Dental School and Hospital), Evaluator: Micro Electronic Applications Ltd., Technical Manager: Ildana Teoranta.

SWB industry partners – 3Com, Astra, Bandwidth Telecommunications Ltd, DCM, DIACOM, Duolog, Educom/Hughes, Europe*Star, iDirect, Intel, MediaSatellite, Netdish/Eutelsat, Satlynx, Visicom Ltd.

In the course of the field trials, a range of issues arose regarding latency, contention ratios and cost, as did new opportunities in the areas of complementary technologies, the impact of which enabled a far more ambitious, exciting and challenging initiative than that originally planned.

Hardware employed in the trials included two-way satelite systems (VSAT and DVB RCS) providing a similar service to ADSI., right up to large-scale systems augmented with wireless local area networks (WLANs) providing town connectivity, and higher end business applications.

Wi-Fi certainly balanced out the scales in terms of cost and deployment (last mile) and enabled the development of a generic model which could be adapted, scaled and tailored to the needs of individual communities/businesses or users groups.

The project went through successful field trials in 2003 and 2004, with a number of locations choosing to retain the satellite (or satellite/WLAN) systems offering a long-term broadband connection to the particular user group (public, business or community/residential). The SWB Final Report is available online at www.swra.ie/broadband.

Finally, an independent evaluation of SWB, its benefits and potential weaknesses is included in Table 16.1.

16.6 Regional broadband aggregation in the UK

The project to be summarized below is based on a specific method called demand aggregation. It is a method for public authorities to solve the needs for broadband communications in the public sector. Among the different models of public participation, it requires the least intervention of the public sector in the implementation and operation of the services, as the public authority is only involved in the collection of demands, aggregating them in order to achieve a critical mass, negotiation with potential providers and in the procurement procedure.

Aggregation refers to bringing total broadband requirements of public sector customers into one procurement requirement to the suppliers. It is accompanied by a network consolidation, by reorganizing several physical networks of customers into a single one.

The UK Government has set an aggressive target of increasing broadband availability from 80 % (at the the end of 2003) to 95 % by 2005 [12]. It was easy to realize that this increase is impossible based on market conditions only, some form of government intervention was necessary. An indirect intervention strategy was chosen based on the aforementioned demand aggregation principle, which is illustrated in Figure 16.6 [13].

The project, launched by the UK Department of Trade and Industry's Broadband Taskforce, is called *the Broadband Aggregation Programme*. It is based on the creation of nine *regional aggregation bodies (trading as Adit)* and one national aggregation body, Adit National. According to the 'Foundation Model', these aggregation bodies act as 'intelligent clients' with

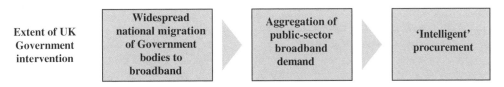

Figure 16.6 The demand aggregation principle

Table 16.1   Independent evaluation of SWB: benefits and potential weaknesses [12]

Benefits of satellite broadband in rural regions	Potential weaknesses and recommendations to address them
• Satellite can quickly provide rural and remote communities with broadband access without significant capital investment. • Satellite should be part of any information society regional development plan and strategy. Satellite can address the broadband needs of sparsely-populated rural regions anywhere, stimulate demand, create competition and establish a broadband market where previously there was none. • eCommunity WLAN/satellite public–private partnerships provide an excellent model to provide broadband to remote communities, but a local champion, commitment and technical planning and support is critical. • Satellite can quickly provide broadband to rural and remote schools (15 % or more of all schools) that otherwise could be waiting for years to receive terrestrial access (if ever). • Satellite can be an intermediate solution (that lights up a whole region with little capital investment). Its low capital cost means that temporary trials can be supported, that support local economic activity, raise awareness of the importance of broadband in the local economy, and hence encourage other suppliers.	• Satellite systems are technically complex. Setting up such systems requires much technical knowledge and support. Reliability/support is critical, or users will lose confidence in the service. *Good preplanning, installation and technical support needs to be specified and resourced at the beginning of any satellite project.* • The greatest benefit of satellite is for remote areas, but as such they are difficult to remotely maintain and technically support. *So the lesson is to keep the system as simple as possible with local basic technical monitoring and support.* • Reliability of satellite services can be a problem due to marginal link budgets or lack of technical support. *Quality of Service needs to be specified, agreed and enforced.* • Satellite supplier organizations can be frustrating and time-consuming, due to their technology focus and lack of user-service orientation. *The satellite industry needs to be more transparent and user-oriented – this has to be addressed by the industry itself.* • Some satellite suppliers promote what is technically possible, but what they can actually make available as a service can fall well short of that (e.g. due to too high contention, or over-promised bandwidth flexibility on demand). *Satellite providers need to be more up-front on what they actually can provide. They need to clarify their offerings in terms of (a) available bandwidth; (b) average speed; (c) contention ratio; (d) interoperability. Satellite providers and the services they provide may need to be regulated.* • Satellite is generally more expensive and outperformed by terrestrial infrastructure such as ADSL (when available). *Be clear on the need for satellite and/or use it as an interim measure to stimulate competition to promote regional development.*

three main roles: as *aggregators, solution developers* and *procurers*. Note that the 'foundation model' covers only the physical connectivity functionality and thus involves interactions with telecommunication infrastructure providers (such as BT, Kingston and Telewest in the UK). The individual Adits can be involved in demand aggregation related to higher layers, namely network integrators and application providers. The tasks of Adits, as outlined in [14], are as follows:

- aggregate public service broadband demands;
- work with public service bodies to coordinate demand aggregation;
- manage the relationship with key stakeholders;
- organize demand such that there is a maximization of value for money and availability;
- use in-house expertise for network design and operations to select solutions that maximize Adit objectives;
- execute the procurement process for tenders using a dedicated 'open' framework contract with 17 OJEU approved suppliers;
- manage the relationship with suppliers;
- monitor SLA performance and compliance.

Regional aggregation bodies are supported by a national body. The Adit National's tasks include managing relationships with key stakeholders at national level, and co-coordinating joint activities for the Adits. Adit National can also act directly for clients who prefer aggregation at national level.

A board comprising its Regional Development Agency and the national Department of Trade and Industry governs each Adit. This ensures the meshing of national and regional policy. Adit National is governed by all of the regional bodies jointly and the DTI.

The governance structure is shown in Figure 16.7.

The project entered its implementation phase at the beginning of 2004, after the strategy was worked out, the Adits were established.

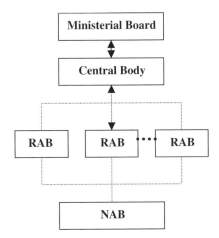

Figure 16.7   The governance structure of regional broadband aggregation

Table 16.2 Comparison of case studies

	Area covered	Technology	Products and Services	Public participation	Customers
Stokab	Stockholm city and county	Fiber	Dark fiber only	100 %	Businesses and public institutions
FastWeb	Milan plus 10+ cities in Italy	Fiber and DSL, IP platform for services	Voice, data, TV, video-on-demand	Minority shares owned by a public utility	Residents
Endesa	Districts of Zaragoza, Barcelona	Power line communication	Internet and telephony	Public utility is a co-owner	Residents and businesses
SWB	South West Ireland region	Satellite	Internet	PPP and EU participation	Residents and businesses
BBA	United Kingdom	Not applicable	Not applicable	Regional and national governments as demand aggregators	Public institutions

The biggest public customer to benefit from the project is the National Health Service (NHS), as the move from the NHSNet to its broadband enhancement is crucial for the implementation of the National Health Programme. The existing network covers the majority of health care providers (hospitals as well as general practitioners). The new network will extend the reach of the existing one by mobile access networks and remote access capabilities to host new applications and services. The high bandwidth will allow for a widespread use of videoconferencing to support telemedicine applications [15].

16.7 Summary

Table 16.2 summarizes the most important aspects of the case studies discussed in this chapter.

Acknowledgments

This chapter was written based on the source materials made kindly available by the management of the projects and institutions/companies involved, to whom the author wishes to express his gratitude:

- Stokab AB;
- e.Biscom S.P.A.;
- Mr Ramiro Alfonsin, Head of Operations and Strategic Planning of the Power Line Communications (PLC) Project of Endesa. Mr Alfonsin is also a member of the board of the PLC Forum [16] and president of the Powerline Utilities Alliance (PUA);
- SWB, an initiative of the South West Regional Authority of Ireland, co-funded by ARTES 4, European Space Agency. The Director of the SWRA is Mr John McAleer;
- Dr Sue Baxter, Director, Broadband Aggregation Programme.

In addition, the author is grateful to Ms Monica Allard Grivans of Stokab, Mr Marina Gillespie of e.Biscom, Mr Fernando Sandoval Cuervo of Endesa and Mr Sinead Crowley of SWRA for kindly revising the text of this chapter and for valuable suggestions.

References

1. Stokab Annual Report, 2003.
2. www.stokab.com
3. Italians Invade Germany, www.boardwatch.com
4. Fiber-optic firm aims to put Milan on the map, *International Herald Tribune*, November 4, 2003, p. 11.
5. Europe Discovers TV via Telephone, *Fortune*, April 19, 2004, p. 19.
6. FastWeb Piles on the Users, www.boardwatch.com
7. PUA (2003), *PLC Current Situation Overview*.
8. Balza, R.A. (2004), *Endesa Net Factory PLC Project*, 2nd World Summit of PLC Associations, Brussels, May 26 2004.
9. www.swra.ie/broadband/content/background.htm
10. *South West Broadband. Exploring Satellite Technologies in Business Education and Public Services*, ESA Workshop, September 2003.
11. www.ildana.ie
12. O'Flaherty, J., *Independent Evaluation of SWB*, Microelectronics Applications Centre Ltd (www.mac.ie).

13. Craine, P. (2003), *UK Broadband Taskforce. A review of the UK approach to increasing the availability of technology neutral broadband access*, Presentation at EU Broadband Workshop, Brussels, December 15, 2003.

14. Hodge, P. (2003), *UK Broadband Aggregation Project*, UK Broadband Task Force, DTI, Presentation at Regional Briefing, June 25, 2003.

15. Arnott, S. (2003), *Broadband Aggregation takes first steps*, July 17, 2003. www.vunet.com/Analysis.

16. www.PLCforum.com

Index

Broadband Services: Business Models and Technologies for Community Networks. Edited by I. Chlamtac,
A. Gumaste and C. Szabó © 2005 John Wiley & Sons, Ltd. ISBN 0-470-02248-5.